Design for Electrical and Computer Engineers

Theory, Concepts, and Practice

Ralph M. Ford
Chris S. Coulston
Penn State University Erie
The Behrend College

D1597047

McGraw-Hill Custom Publishing

Boston Burr Ridge, IL Dubuque, IA Madison, WI New York San Francisco St. Louis
Bangkok Bogotá Caracas Lisbon London Madrid
Mexico City Milan New Delhi Seoul Singapore Sydney Taipei Toronto

The McGraw·Hill Companies

Design For Electrical and Computer Engineers:
Theory, Concepts, and Practice

McGraw-Hill's Custom Publishing consists of products that are produced from camera-ready copy. Peer review, class testing, and accuracy are primarily the responsibility of the author(s).

1 2 3 4 5 6 7 8 9 0 QSR QSR 0 9 8 7 6 5

ISBN 0-07-319599-5

Editor: Julie Kehrwald
Production Editor: Nina Meyer
Printer/Binder: Quebecor World

Preface

This book is written for undergraduate students and teachers engaged in electrical and computer engineering (ECE) design projects, primarily in the senior year. The objective of the text is to provide a treatment of the design process in ECE with a sound academic basis that is integrated with practical application. This combination is necessary in design projects because students are expected to apply their theoretical knowledge to bring useful systems to reality. This topical integration is reflected in the subtitle of the book: Theory, Concepts, and Practice. Fundamental theories are developed whenever possible, such as in the chapters on functional design decomposition, system behavior, and design for reliability. Many aspects of the design process are based upon time-tested concepts that represent the generalization of successful practices and experience. These concepts are embodied in processes presented in the book, for example, in the chapters on needs identification and requirements development. Regardless of the topic, the goal is to apply the material to practical problems and design projects. Overall, we believe that this text is unique in providing a comprehensive design treatment for ECE, something that is sorely missing in the field. We hope that it will fill an important need as capstone design projects continue to grow in importance in engineering education.

We have found that there are three important pieces to completing a successful design project. The first is an understanding of the design process, the second is an understanding of how to apply technical design tools, and the third is successful application of professional skills. Design teams that effectively synthesize all three tend to be far more successful than those that don't. The book is organized into three parts that support each of these areas.

The first part of the book, the *Design Process*, embodies the steps required to take an idea from concept to successful design. At first, many students consider the design process to be obvious. Yet it is clear that failure to understand and follow a structured design process often leads to problems in development, if not outright failure. The design process is a theme that is woven throughout the text; however, its main emphasis is placed in the first four chapters. Chapter 1 is an introduction to design processes in different ECE application domains. Chapter 2 provides guidance on how to select projects and assess the needs of the customer or user. Depending upon how the design experience is structured, both students and faculty may be faced with the task of selecting the project concept. Further, one of the important issues in the engineering design is to understand that systems are developed for use by an end-user, and if not designed to properly meet that need, they will likely fail. Chapter 3 explains how to develop the Requirements Specification along with methods for developing and documenting the requirements. Practical examples are provided to illustrate these methods and techniques. Chapter 4 presents concept generation and evaluation. A hallmark of design is that there are many potential solutions to the problem. Designers need to creatively explore the space of possible solutions and apply judgment to select the best one from the competing alternatives.

The second part of the book, *Design Tools*, presents important technical tools that ECE designers often draw upon. Chapter 5 emphasizes system engineering concepts including the well known functional decomposition design technique and applications in a number of ECE problem domains. Chapter 6 provides methods for describing system behavior, such as flowcharts, state diagrams, data flow diagrams and a brief overview of the Unified Modeling Language (UML). Chapter 7 covers important issues in testing and provides different viewpoints on testing throughout the development cycle. Chapter 8 addresses reliability theory in design, and reliability at both the component and system level is considered.

The third part of the book focuses on *Professional Skills*. Designing, building, and testing a system is a process that challenges the best teams, and requires good communication and project management skills. Chapter 9 provides guidance for effective teamwork. It provides an overview of pertinent research on teaming and distills it into a set of heuristics. Chapter 10 presents traditional elements of project planning, such as the work breakdown structure, network diagrams, and critical path estimation. It also addresses how to estimate manpower needs for a design project. Chapter 11 addresses ethical considerations in both system design and professional practice. Case studies for ECE scenarios are examined and analyzed using the IEEE (Institute of Electrical and Electronics Engineers) Code of Ethics as a basis. The book concludes with Chapter 12, which contains guidance for students preparing for oral presentations, often a part of capstone design projects.

Features of the Book

This book aims to guide students and faculty through the steps necessary for the successful execution of design projects. Some of the features are listed below.

- Each chapter provides a brief motivation for the material in the chapter followed by specific learning objectives.

- There are many examples throughout the book that demonstrate the application of the material.

- Each end-of-chapter problem has a different intention. Review problems demonstrate comprehension of the material in the chapter. Application problems require the solution of problems based upon the material learned in the chapter.

Design problems are directly applicable to design projects and are usually tied in with the Project Application section.

- Nearly all chapters contain a Project Application section that describes how to apply the material to a design project.
- **Some chapters contain a Guidance section that represents the author's advice on** application of the material to a design project.
- Checklists are provided for helping students assess their work.
- There are many terms used in design whose meaning needs to be understood. The text contains a glossary with definitions of design terminology. The terms defined in the glossary (Appendix A) are indicated by *italicized-bold* highlighting in the text.
- All chapters conclude with a Summary and Further Reading section. The aim of the Further Reading portion is to provide pointers for those who want to delve deeper into the material presented.
- The book is structured to help programs demonstrate that they are meeting the ABET (accreditation board for engineering programs) accreditation criteria. It provides examples of how to address constraints and standards that must be considered in design projects. Furthermore, many of the professional skills topics, such as teamwork, ethics, and oral presentation ability, are directly related to the ABET Educational Outcomes. The requirements development methods presented in Chapter 3 are valuable tools for helping students perform on cross-functional teams where they must communicate with non-engineers.
- An instructor's manual is available. It contains not only solutions, but guidance from the authors on teaching the material and managing student design teams. It is particularly important to provide advice to instructors since teaching design has unique challenges that are different than teaching engineering science oriented courses that most faculty are familiar with.

How to Use this Book

There are several common models for teaching capstone design, and this book has the flexibility to serve different needs. Particularly, chapters from the Professional Skills section can be inserted as appropriate throughout the course. Recommended usage of the book for three different models of teaching a capstone design course is presented.

- **Model I.** This is a two-semester course sequence. In the first semester, students learn about design principles and start their capstone projects. This is the model that we follow. In the first semester the material in the book is covered in its entirety. The order of coverage is typically Chapters 1–3, 9, 4–6, 10–11, and 7–8. Chapter 9 (Teams and Teamwork) is covered immediately after the projects are identified and the teams are formed. Chapters 10 (Project Management) and 11 (Ethical and Legal Issues) are covered after the system design techniques in Chapters 5 and 6 are presented. Students are in a good position to create a project plan and address ethical issues in their designs after learning the more technical aspects of design. Chapter 12 (Oral Presentations) is assigned to students to read before their first oral presentation to the faculty. The course concludes with principles of testing and system reliability (Chapter 7 and 8). We assign a good number of end-of-chapter problems and have quizzes throughout the semester. By the end of the first semester, teams are expected to have completed development of the requirements, the high-level or architectural design, and developed a project plan. In the second semester, student teams implement and test their designs under the guidance of a faculty advisor.

- **Model II**. This two-semester course sequence is similar to Model I with the difference being that the first semester is a lower credit course (often one credit) taught in a seminar format. In this model chapters can be selected to support the projects. Some of the core chapters for consideration are Chapters 1–5, which take the student from project selection to functional design, and Chapters 9–11 on teamwork, project management, and ethical issues. Other chapters could be covered at the instructor's discretion. The use of end-of-chapter problems would be limited, but the project application sections and example problems in the text would be useful in guiding students through their projects.

- **Model III**. This is a one-semester design sequence. Here, the book would be used to guide students through the design process. Chapters for consideration are 1–5 and 9–10, which provide the basics of design, teamwork, and project management. The project application sections and problems could be used as guidance for the project teams.

Acknowledgements

Undertaking this work has been a challenging experience and could not have been done without the support of many others. First, we thank our families for their support and patience. They have endured many hours and late evenings that we spent researching and writing. Melanie Ford is to be thanked for her diligent proofreading efforts. Bob Simoneau, the Director of the Penn State Behrend School of Engineering and Engineering Technology, has been a great supporter of the book and has also lent his time in reading and providing comments. Our school has a strong design culture, and this book would not have happened without that emphasis; our faculty colleagues need to be recognized for developing that culture. Jana Goodrich and Rob Weissbach are two faculty members with whom we have collaborated on other courses and projects. They have influenced our thinking in this book, particularly in regard to project selection, requirements development, cost estimation, and teamwork. We must also recognize the great collaborative working environment that exists at Penn State Behrend, which has allowed this work to flourish. Our students have been patient in allowing us to experiment with different material in the class and on the projects. Examples of their work are included in the book. In particular, we would like to thank former students Jeff Betts, Jon Ciccozzi, Thad Esek, Roger Grayson, John Hunt, Stephen Haug, Dale Lamb, Nate Lewis, Dante Vilairoj, and William Wykoff for their contributions. John Wallberg contributed the disk drive diagnostics case study in Chapter 11 that we have found very useful for in-class discussions. John developed this while he was a student at MIT. Thanks to Anne Maloney for her copyediting of the manuscript. Finally, Carlise Stembridge and Julie Kehrwald at McGraw-Hill have been very supportive and we thank them for their efforts to make this book a reality.

We hope that you find this book valuable, and that it motivates you to create great designs. We welcome your comments and input. Please feel free to email us.

Ralph M. Ford, rmf7@psu.edu
Chris S. Coulston, csc104@psu.edu

Table of Contents

Part I – The Engineering Design Process

Part II – Design Tools

Part III – Professional Skills

Appendices

Part I – The Engineering Design Process

Chapter 1   The Engineering Design Process

en-gi-neer (n) 1. *One versed in the design, construction, and use of machines.* 2. *One who employs the innovative and methodical application of scientific knowledge and technology to produce a device, system, or process, which is intended to satisfy human needs.*—American College Dictionary

Take a moment to read and analyze the key elements of the two definitions presented above. If you are an engineering student or practicing engineer, do you think that this definition applies to you? The first definition uses the terms *design* and *construction*. People like to think of themselves as designers. Why is that so? The answer may be in the combination of the term *construction,* and from the second definition, the idea of *innovation.* Applying innovation and creativity to produce something new is a wonderfully rewarding process. The great thing about being an engineer is that it allows you to be a creative designer. That is generally not the way the profession is viewed. What is the difference between engineering design and other types of design that are associated with creativity such as interior design, fashion design, or webpage design? The answer is supplied in the second definition which states "...*methodical application of scientific knowledge and technology*..." As an engineering student, you have studied a great deal of math, science, and fundamental technology, but probably have had limited exposure to creative and innovative design.

The definition also contains the somewhat contradictory terms *innovative* and *methodical.* If there is an established and methodical way of employing a scientific principle or process, it does not seem to allow much room for creativity and innovation. The truth is that the two concepts are in competition with each other, but a good engineer realizes this and utilizes both effectively. The definition also indicates that engineers design to satisfy human needs, an important, yet often overlooked point. That means that when designing systems, it is necessary to consider the needs of the user and ethical application of the technology.

This book aims to help electrical and computer engineers become effective designers, to better understand professional practices, and to provide guidance for executing design projects. This chapter presents the processes by which designs are realized, the characteristics of successful engineers, and an overview of the book.

Learning Objectives

By the end of this chapter, the reader should:

- Understand what is meant by engineering design.
- Understand the phases of the engineering design process.
- Be familiar with the attributes of successful engineers.
- Understand the objectives of this book.

1.1 The Engineering Design Process

ABET (formerly known as Accreditation Board of Engineering and Technology) provides the following definition of engineering design [ABE03].

> *Engineering design is the process of devising a system, component, or process to meet desired needs. It is a decision-making process (often iterative), in which the basic sciences, mathematics, and engineering sciences are applied to convert resources optimally to meet a stated objective. Among the fundamental elements of the design process are the establishment of objectives and criteria, synthesis, analysis, construction, testing, and evaluation.*

The definition indicates that, in engineering design, different phases of the process have to be re-visited and the deliverables for the phase updated as necessary. Realistic problems are complex with many potential solutions; the goal is not to find just any solution, but the best one given the constraints and available resources. This requires the application of sound judgment, decision-making skills, and patience in constantly evaluating progress towards a solution. The definition identifies some common elements of the design process, such as establishment of criteria, synthesis, construction, and testing.

Design processes embody the steps required to take an idea from concept to realization of the final system, and are problem-solving methodologies that aim to develop a system that best meets the customer's need within given constraints. This is not all that different from some everyday processes, such as preparing dinner. Say you are hungry and need to eat dinner before you can go to see a movie that starts in one hour. The constraints are time, money, food, your tastes, and nutritional value if you are health-conscious. You brainstorm and come up with the options of making dinner at home, going to a restaurant, or buying something to eat at the theater. Based on these options, you then select the solution based on your evaluation of the best one. This is similar in philosophy to the stages of design processes where you have a problem to solve, constraints, and a number of potential solutions to select from.

A related term is known as the *product realization process*. The product realization process is broader in scope, including aspects such as entrepreneurship, market research, financial planning, product pricing, and market strategy. Many technologies have their own particular design processes that have evolved over time and have been found by practitioners in the field to be valuable. For example, different methodologies are applied in the design of integrated circuits (VLSI), embedded systems, and software systems, yet they all have some degree of commonality, such as requirements analysis, technical design, and system test. Design processes continue to evolve. One field in which this is particularly true is in software design due to the constantly changing nature of software and the special challenges that large software projects pose.

Cross [Cro00] identified two types of design processes—prescriptive and descriptive. As the name implies, ***prescriptive design processes*** set down an exact process, or systematic recipe, for realizing a system. Prescriptive design processes are often algorithmic in nature and expressed using flow charts with decision logic. An example of a prescriptive process is shown in Figure 1.1, which describes the front end of the design process where the problem and requirements are determined. A decision block is included where the requirements are examined to determine if they satisfy the needs of the problem. ***Descriptive processes*** are less formal, describing typical activities involved in realizing designs with less emphasis on exact sequencing. The distinction between descriptive and prescriptive processes is not always clear, however, and some processes may be considered more strongly associated with one property than the other. Cross makes an important point in stating that design processes are sometimes viewed as common sense and thus ignored, resulting in failed products. Cross cites two good reasons to adhere to design processes: 1) they formalize thought processes to ensure good practices are followed, leading to better and more innovative solutions, and 2) they keep all members of the team synchronized in terms of understanding where they are in the design process.

Figure 1.1 A prescriptive design process for problem identification and requirements selection.

A descriptive process that is widely applicable to design problems is shown in Figure 1.2. In a perfect world, the process starts with the identification of the problem, proceeds clockwise to research, followed the specification phase, and so on until the system or device is delivered and goes into service (maintenance phase). This scenario is unrealistic, ignoring the iterative nature of design where the design team alternates between different phases as necessary. Consequently, links are inserted that allow transitions between all the different phases of the

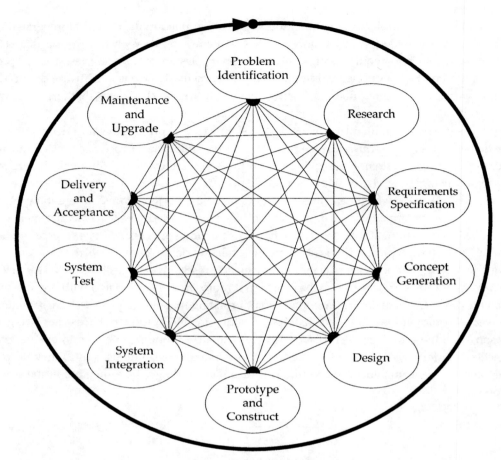

Figure 1.2 A descriptive overview of the design process.

design process. Of course, transitions between certain phases are unreasonable or very costly. It is virtually impossible to move directly from problem identification to system integration without developing a design concept first. It is much more likely for engineers to alternate between nearby phases in the process, such as problem identification, research, requirements specification, and concept generation. This does not mean that you can't move between phases that are not in close proximity in the model. For instance, the customer's needs may change while in the design phase, necessitating re-evaluation of the needs, correction of the requirements specification, and system redesign—all at a substantial cost in time and money. Studies have shown that the cost required to correct errors or make changes increases exponentially as the project lifetime increases, as presented in Figure 1.3.

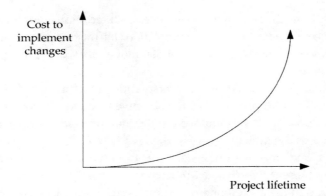

Figure 1.3 The cost to implement design changes which increases exponentially as project lifetime increases.

1.1.1 Elements of the Design Process

Nearly all the phases of the design process in Figure 1.2 are covered in this book, with the exception of the maintenance phase. The objective of the first phase, *problem identification*, is to identify the problem and customer needs. This occurs in a variety of ways, from someone conceiving a new idea to a client coming to you with a problem to solve. In either case, it is important to determine the true needs for the product, device, or system (terms that are used interchangeably throughout the book and often referred to as systems). Failure to correctly identify the needs has negative ramifications for the entire process, typically resulting in costly redesigns, or even worse, abandonment of the project.

In the *research phase* the design team conducts research on the basic engineering and scientific principles, related technologies, and existing solutions. The objective is to become experts on the particular technology, save time and money by not re-inventing the wheel, and be positioned to develop new and innovative solutions.

The *Requirements Specification* articulates what the system must do for it to be successful and to gain acceptance by the customer. It is important to focus on what the system must do, as opposed to how the solution will be implemented. This is challenging since engineers tend to focus on solutions and propose implementations early in the process. This is not surprising since engineering education focuses on solving problems rather than specifying them. The requirements are the mission statement that guides the entire project, and if properly developed, provide flexibility for creativity and innovation in developing solutions.

In *concept generation*, many possible solutions to the problem are developed. The hallmark of design is that it is open-ended, meaning that there are multiple solutions to the problem and the objective is to develop the one that best meets the requirements and satisfies the constraints. In this phase, wild creativity is encouraged, but it is ultimately tempered with critical evaluation of the competing alternatives.

In the *design phase*, the team iteratively develops a technical solution, ultimately producing a detailed system design. Upon its completion, all major systems and subsystems are identified and described using an appropriate model that depends upon the particular technology being employed.

In the *prototyping and construction phase*, different elements of the system are constructed and tested. In rapid prototyping, the objective is to model some aspect of the system, demonstrating functionality to be employed in the final realization. Many prototypes are discarded or modified as the system evolves—the idea is to experiment, demonstrate proof-of-concept principles, and improve understanding. Prototypes may be used anywhere in the process—you may present the client with prototypes after the concept generation phase, or they may be utilized in the design phase to test a design idea, or as the final system is tested and developed.

During *system integration*, all of the subsystems are brought together to produce a complete working system. This phase is challenging and time-consuming since many different pieces of the design must be interfaced, and the team must work closely to make it all work. Care taken in the design phase to clearly communicate the functionality and interfaces between subsystems will aid in system integration. System integration is closely tied to the *test phase*, where the overall system is tested to demonstrate that it meets the requirements.

Ultimately the system is *delivered* to the customer where it is likely that they will test it using a mutually agreed upon process. Development does not necessarily end when the system goes into service, as it will likely enter the *maintenance phase* where it is maintained, upgraded to add new functionality, or where design problems are corrected. Following and understanding the design process improves the probability of successful system development. The process is flexible, and the designer needs to transition between different phases in order to bring the system to realization. Design is an iterative process, you may not fully understand everything necessary in any given phase and have to revisit different steps as the system evolves. That is not a license for not trying to develop the best design you can on the first attempt—by all means do so—but realize that flexibility and a willingness to change the design are necessary.

1.1.2 Technology Specific Design Processes

Different application domains have developed specialized processes for technology-specific design. One such example is VLSI (Very Large Scale Integration) design. A typical VLSI design process is shown in Figure 1.4 [Wol02]. In this model the system specification is used to develop the system architecture. The system architecture is composed of the major functional units that constitute an integrated circuit. Each functional unit is then designed at the gate logic level, which is subsequently designed at the circuit (transistor) level, and finally the circuit elements are laid out on the silicon chip. This is an excellent demonstration of the divide-

and-conquer approach to design, where a complex system is broken down into lower levels of abstraction and each of these is further broken down until the design objectives are met.

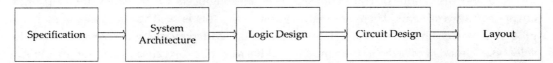

Figure 1.4 A process for integrated circuit (VLSI) design [Wol02].

Next, consider the design process for embedded computer systems shown in Figure 1.5. Embedded systems are combined hardware/software systems embedded into a larger system to perform dedicated application specific operations. Embedded systems are employed in automobiles, DVD players, and digital cameras to name a few applications. Performance issues dominate embedded applications, and the designer needs to partition tasks between software and hardware to achieve optimum performance. This design process is somewhat prescriptive with phases for requirements gathering, specifications, and architectural design. The process reflects the unique nature of embedded systems with separate software and hardware design blocks, married together by the interface design.

Figure 1.5 An embedded system design process [Ern97].

The field of software engineering is one in which the development of different design process models is still under considerable flux today. This is due to the complex nature of software and the failure of computer scientists and engineers to effectively develop high-quality software systems. There are many reasons why this is so. The sheer size of software programs may easily exceed one million lines of code written by many different software developers. One small mistake in those millions of lines of code can cause the system to fail. Another difficulty is in designing for upgrade and reuse of software. What if the needs change after the millions of lines of code are developed and one of the fundamental structures or objects needs to be upgraded?

The *waterfall model* shown in Figure 1.6 is one of the first proposed and most well-known software design processes. In this model the development proceeds linearly from the first step where the user's needs are analyzed through the phases of specification development, design, test, and maintenance. This works well for well-defined and moderately complex software applications, but fails as complexity grows due to the inability to move between phases. A more flexible software design process is known as the *spiral model*, which is a cyclical process where phases are revisited as necessary [Som01]. *Extreme Programming* is a more recent and controversial software development process, where relatively small teams of software developers rapidly develop software following some strict rules. Both the spiral model and Extreme Programming are examined in more detail in the end of chapter problems.

Figure 1.6 Waterfall software development process. In this model, development proceeds linearly from requirements analysis, through each subsequent phase, terminating with maintenance.

1.2 The World-Class Engineer

The ability to effectively design is important for engineers, requiring strong technical skills and an understanding of the design process. Yet, this ability in itself is not enough to become an effective practicing engineer. The Pennsylvania State University Leonhard Center for the Advancement of Engineering Education, in consultation with a number of industries, developed a description of what is referred to as a "World-Class Engineer" [Leo95]. Shown in Table 1.1, the description identifies the characteristics of successful engineers, and contains six major elements: 1) Aware of the World, 2) Solidly Grounded, 3) Technically Broad, 4) Effective in Group Operations, 5) Versatile, and 6) Customer-Oriented. A few key aspects of the description are discussed here. It recognizes that engineers must be effective in group operations, since the majority of projects are carried out in teams. Not only that, many projects span multiple technical disciplines and are executed in multifunctional organizations that have diverse groups such as marketing, finance, human resources, technical support, and service. It also recognizes that an engineer must be versatile, innovative, understand ethical principles, and be customer-oriented, important themes that are stressed throughout this book.

1.3 Book Overview

Consider the digital camera, the cellular phone, and the space shuttle, all complex systems that integrate a variety of technologies. A digital camera is the synthesis of an embedded electronics system: optics, a mechanical lens assembly, and the camera package itself. The embedded electronics contain an imaging integrated circuit, a digital display, digital interface circuitry, flash memory storage, system control software, and the user interface. The challenges of integrating the components of such a system and having it record and transfer huge amounts of image data, within an acceptable timeframe, are immense. Cellular phones are another good example of a complex system that represents a technology that has shrunk in size, but increased tremendously in functionality at the same time. They encompass digital data communications, antenna design, encryption for secure data transmission, a user interface display, and Internet connectivity. At the other end of the spectrum are large-scale space and military systems, such as the space shuttle. Despite the two shuttle accidents, the safety and reliability requirements of the space shuttle are incredibly high. Realizing such a system is accomplished by a tremendous number of people from many disciplines working for different organizations. All three of these technologies were developed by large teams that encompass multiple disciplines. The processes and practices employed in their development represent application of the fundamentals that this book hopes to cover. While you won't be building complete space shuttles by the end of this design course, you can expect to apply design principles that allow you to design and integrate a relatively complex system, maybe even a part of the space shuttle.

Table 1.1 The World-Class Engineer (Copyright the Leonhard Center for the Advancement of Engineering Education, The Pennsylvania State University. Reprinted by permission.)

I. Aware of the World
- sensitive to cultural differences, environmental concerns, and ethical principles
- alert to market opportunities (both high- and low-tech)
- cognizant of competitive talents, work ethic, and motivation

II. Solidly Grounded
- thoroughly trained in the fundamentals of a selected engineering discipline
- has a historical perspective and remains aware of advances in science that can impact engineering
- realizes that knowledge doubles at breakneck speed and is prepared to continue learning throughout a career

III. Technically Broad
- understands that real-life problems are multidisciplinary
- thinks broadly, seeing an issue in a rich context of various alternatives, probabilities, etc., rather than a narrow quest to find a single answer
- is conversant in several disciplines
- is trained in systems modeling and the identification of critical elements. Understands the need to design experiments to verify or extend analysis, as well as meet specification requirements
- is psychologically prepared to embrace any field necessary to solve the problem at hand

IV. Effective in Group Operations
- cooperative in an organization of individuals working toward a common creative goal that is often multidisciplinary and multifunctional in nature
- effective in written and oral communication
- willing to seek and use expert advice
- cognizant of the value of time and the need to make efficient use of the time in all phases of an endeavor
- understanding and respectful of the many facets of business operation -- general management, marketing, finance, law, human resources, manufacturing, service, and especially quality

V. Versatile
- innovative in the development of products and services
- sees engineering as applicable to problem solving in general
- considers applying engineering beyond the typical employment focus of engineering graduates in the manufacturing industries, to the much broader economy (financial services, health care, transportation, etc.) where engineering skills could make a dramatic improvement in the productivity of those segments of the economy that employ 80 percent of the U.S. population

VI. Customer Oriented
- realizes that finding and satisfying customers is the only guarantee of business success
- understands that products and services must excel in the test of cost-effectiveness in the global marketplace

Figure 1.7 The guiding philosophy of this book. In order to achieve success in executing engineering and design projects, it takes an understanding of the design process, strong technical design tools, and professional skills.

The intention is to teach the application of design principles to computer and electrical engineers and to help prepare students for a professional career. The majority of engineering education is devoted to math, science, engineering science, and problem-solving. They are important topics required to enter this highly technical field. However, it is clear that there are other aspects beyond this that are equally important for success, including an understanding of system design, innovation, ethical principles, teamwork, and strong communication skills.

The book is divided into three parts: I–Design Process, II–Design Tools, and III–Professional Skills. This is shown in Figure 1.7 as three separate, but related components that play a key role in achieving Project Excellence—the ability to complete a project, in an ethical manner that meets the customer's need, satisfies the constraints, and is clearly communicated to all involved. The chapters are decoupled as much as possible so that the reader can move between chapters as necessary. In Part I, the emphasis is on understanding and gaining experience in the different phases of the design process. The reader is guided through the steps of project identification, research, specification development, creative concept generation, and critical evaluation of competing solutions. Part II addresses topics that are often employed in design, including functional decomposition, description of system behavior, reliability, and testing. Part III addresses professional skills, including teamwork principles, project planning, ethics in design and the profession, and oral communication skills.

Here are a few thoughts to conclude the chapter and get started on the path to great designs. You are embarking on what will likely be a fun, challenging, sometimes frustrating, and ultimately rewarding journey. The systems that electrical and computers engineers work with continue to become increasingly complex and multi-disciplinary in nature. The example problems presented in this book come from the fields of analog electronics, digital electronics, electrical systems theory, and software systems. These four areas comprise a significant problem-domain common to the education of most electrical and computer engineers. Finally, consider the quote below by Robert Hayes on the importance of design.

> *Fifteen years ago, companies competed on price. Today it's quality. Tomorrow it's design.* —
> Robert H. Hayes, Harvard Business School, 1991.

What is this saying? Well, it is clear that the world continues to move to a more knowledge-based society, where individuals and companies compete on the strength of their intellectual capital and ability to produce new and innovative products. That is what design is all about. It is not saying that price and quality are unimportant, they certainly are; in fact quality and reliability in design are part of this book. It is that quality and price are a given, and successful products will be distinguished by their design characteristics. The implication is that design will play a larger role in the development and success of products. The future Hayes predicted is now. Design is what will distinguish between products that are seen as commodities and those that are truly unique and profitable.

1.4 Summary and Further Reading

Engineering design is an iterative process in which the design team employs creativity and technical knowledge to develop a solution that best meets the end-users' needs within the constraints applied to the problem. There is no single design process that can be applied to all situations and technologies, but there are many common elements shared, regardless of the technology under consideration. In order to successfully bring designs to fruition, it takes a combination of design tools, professional skills, and a clear understanding of the process needed to complete designs. The objective of this book is to develop your proficiency in these areas so that you may become an effective engineer and achieve excellence in design projects.

Engineering Design Methods by Nigel Cross [Cro00] presents the differences between descriptive and prescriptive design processes, and covers a wide array of processes in more detail. It also discusses the cognitive characteristics of effective designers. There are many good books on software engineering process development methods. Software Engineering by Ian Sommerville [Som01] discusses the different software design process models, such as the waterfall and spiral models. This is also true of many modern software engineering texts. The original reference to the waterfall model is by Royce [Roy70]. The Art of Innovation by Michael Kelley [Kel01] describes the activities of well-known design company IDEO and is a highly readable description of their design practices. The ABC *Nightline* news program also produced an interesting segment on IDEO [ABC01] that can be purchased at the ABC website. The Circle of Innovation by Tom Peters [Pet97] is another popular book that provides his perspective on current trends in business and the importance of design.

1.5 Problems

1.1. In your own words, describe the difference between prescriptive and descriptive design processes. Cite examples of each.

1.2. Describe the relationship between the Problem Identification, Research, and Requirements Specification phases of the design process.

1.3. Describe the relationship between the Concept Generation and Design phases of the design process.

1.4. Construct a prescriptive design process for the Problem Identification, Research, Specification, Concept, and Design phases of the design process. The result should be a flow chart that contains decision blocks and iteration as necessary.

1.5. Describe the main differences between the VLSI and embedded system design processes.

1.6. Using the library or Internet, conduct research on the spiral software design process. What are the advantages and disadvantages of this relative to the waterfall model? Be sure to cite references.

1.7. Using the library or Internet, conduct research on the Extreme Programming design process. Outline the significant elements of the Extreme Programming paradigm. What are the pro and con arguments for this software development model? Be sure to cite references.

1.8. **Project Application.** In preparation for the project and team selection, develop a personal inventory that includes a list of five favorite technologies or engineering subjects that you are interested in pursuing. Also, list the strengths and weaknesses that you bring to a project team.

Chapter 2 Project Selection and Needs Identification

For every problem there is a solution that is simple, neat, and wrong.—H.L. Mencken

Business, industry, and organizations are shifting away from a management, or function-oriented, structure and becoming project focused. Traditionally, companies have organized resources based on functions such as accounting, engineering, finance, manufacturing, and marketing. It is often more effective to organize around projects that are of significant value and align resources to meet the needs of the project. This means that traditional departments and middle management are being de-emphasized and the role of projects is growing. Capstone design projects provide a great opportunity to gain experience in the management and execution of a project. One of the first and most important decisions encountered is selecting the project concept.

The focus of this chapter is to provide pragmatic guidance in the project selection phase. A description of design and engineering projects is presented, followed by advice on how projects should be selected by organizations and engineering students who wish to put design principles into practice. The chapter addresses how to identify the needs of the end-user and provides guidance for conducting background research. All of this information is brought together in a project proposal that identifies the needs being addressed, the goals of the project, and research on the technology.

Learning Objectives

By the end of this chapter, the reader should:

- Have an understanding of the types of projects that electrical and computer engineers undertake.
- Understand and be able to apply sound criteria for project selection.
- Know how to determine, document, and rank end-user needs.
- Be familiar with some of the resources available for conducting research on design projects.
- Have selected a project concept and developed a Problem Statement.

2.1 Engineering Design Projects

This section provides a classification of design and describes some of the types of projects undertaken by practicing engineers and those tackled in student projects. In reality, most projects don't fit neatly into the categories presented, but are some combination of them. The objective of a design project is to create a new *artifact* (system, component, or process) to meet a given need. Within the design domain there are different types of designs that are classified broadly into three categories of Creative, Routine, and Variant designs [Cro00].

Creative designs represent new and innovative products. An example of a creative design is the PalmPilot™ Personal Digital Assistant (PDA). While the idea for the PDA had been around for awhile, earlier attempts at developing the technology, notably the Apple Newton, were unsuccessful. This was primarily due to unreliable handwriting recognition that frustrated the user. However, Palm Computing had the creative idea to develop a simplified handwriting language, Graffiti, which eliminated the need for natural handwriting recognition. The PalmPilot is a great example of a creative design—it is simple (four basic functions), fits in your pocket, and is easy to use. This innovation spawned a huge hand-held computing industry.

Variant designs are variations of existing designs, where the intent is to improve performance or add features to an existing system. Many engineering projects fall into this category. For example, the objective may be to increase accuracy or system throughput.

Routine designs represent the design of devices for which theory and practice are well-developed. Examples are DC power supplies, analog and digital filters, and basic digital components such as adders and comparators. Routine designs are often components of more complex creative and variant designs.

Within these three categories of design, there are many different types of projects. *Systems engineering and systems integration projects* represent the synthesis of many subsystems into a larger system. They may be creative or variant designs, but have unique challenges since they are typically large and involve many people and technologies. Adherence to good design processes is important for their success. Engineers are often engaged in *systems test*, where the objective is to ensure that a system meets stated requirements and the needs of the user. Examples include the testing of systems for use in space and military environments.

The objective in *experimental design projects* is to design experimental procedures and apparatus for determining the characteristics of a system. For example, an engineering team may test a system under a variety of operating conditions. The Problem Statement in Example 2.2 (Section 2.5) presented later in the chapter is such a project. The objective was to design a series of experiments to test the feasibility of gigabit Ethernet technology in a military environment. The test explored the impact of environmental factors such as temperature and vibration, and further used this data to estimate the operating lifetime of the Ethernet board. Upon completion of this project, the team made recommendations as to the allowable operating ranges of the technology.

The objective in *analysis projects* is to analyze some aspect of an existing system to improve or correct it. For example, a system or process may be failing in the field and the source of the failure unknown. Tools such as the Failure Mode Effects and Analysis technique may be applied in this situation to identify the sources of failure. In *technology evaluation projects*, technologies are assessed to determine if they can be used in a given application. This may be to determine if the technology can improve an existing system, or to characterize its operating performance.

The objective of a *research project* is to perform research or experiments with the goal of discovering or creating a new technology. The fundamental difference between this and other types of projects is that the ultimate outcomes are unknown. Most engineering research falls under the category of *applied research*. This refers to the creation of new technology or systems based on existing technology and theory developed from fundamental research. *Fundamental research* emphasizes the discovery of new scientific principles without necessarily having an intended application. Fundamental research is very valuable, but not typically a part of design projects.

2.2 Project Feasibility and Selection Criteria

Selecting a project can be trying since the objectives are unknown, the technologies are not defined, and research on the concept needs to be done. This section provides questions to ask and issues to consider in the early stages of project selection. George H. Heilmeier (an electrical engineer who has held positions as Chief Technology Officer of Texas Instruments, Director of the Defense Advanced Research Projects Agency, and CEO of Bellcore) developed a set of questions to answer when starting a new project [Sha94]. Heilmeier argued that all projects must be tied to the goals of the organization, and applied this by asking the following questions:

- What are you trying to do? Articulate your goals using absolutely no jargon.
- How is it done today, and what are the limitations of current practice?
- What is new in your approach, and why do you think it will be successful?
- Who cares? If you are successful, what difference will it make?
- What are the risks and payoffs?
- How much will it cost? How long will it take?
- What are the midterm and final exams to check for success?

Heilmeier credits successful completion of projects that he managed to answering these questions up-front and adhering to disciplined project management processes.

A second perspective is offered from an organizational project management viewpoint [Gra02] that provides the following criteria for project selection:

- *The project must be tied to the mission and vision of the organization.* Believe it or not, organizations often spend resources fruitlessly on projects that don't meet this criterion. To be fair, there is always risk associated with a project and it is sometimes hard to judge exactly how well a project meets this criterion. Neglecting the mission and vision sometimes results from a failure to ask fundamental questions about the project. For engineers who are new to an organization, it is hard to judge a project's importance relative to the mission and goals, but if you find yourself in this situation, do not be afraid to ask some questions. Novices ask basic questions that are often overlooked by those who are highly experienced or intimately involved in a project.

- *Must have payback.* An economic analysis should be done to estimate if the project will make a profit. Much of this is outside the scope of this text, requiring marketing and financial analyses. Chapter 10 covers the basics of project cost estimation that will help in trying to answer this question.

- *Should have Selection Criteria.* Sound criteria for selecting among competing projects should be employed. The needs of the customer must be understood and prioritized, and based upon this, criteria developed for comparing competing project concepts. This ranking can be quantified as shown in the decision matrix in Table 2.1, where each concept is scored against pre-selected criteria. In this chapter, strategies for determining and ranking customer needs are examined, while Chapter 4 explores the application of decision matrices in more detail.

Table 2.1 Hypothetical decision matrix for comparing project concepts. A 1-10 scale for rating the concept vs. the criteria is used.

Selection Criteria	Concept 1	Concept 2	Concept 3
Profitability	5	4	5
Market Share Potential	3	8	7
Projected Development Time	7	6	8
Score	15	18	20

- *Objectives of the Project should be SMART: Specific, Measurable, Assignable, Realistic, Time-Related.* Chapter 3 addresses how to determine project requirements that are Specific and Measurable. Assignable, Realistic, and Time-Related all refer to project management aspects that are covered in Chapter 10. The objective is to develop tasks that are assigned to groups or individuals and realistically can be completed in the given timeframe.

When selecting a capstone project, many of these questions and criteria apply. When proposing the project idea, the problem should be clearly stated, the limits of current practice researched, and the significance of the project understood. Criteria for selecting a project based upon your particular environment can be identified and used to compare the concepts. The objectives of the project should meet the SMART criterion. This allows the tasks to be divided between the team members, responsibilities assigned, and a time-frame for completion identified.

2.3 Needs Identification

Often a customer, client, or supervisor comes to you with a problem to solve and you must determine the needs or requirements for the solution to the problem. In other words, determine the *voice of the customer*. This seems like a simple statement—ask the customer what they want and you are done, right?

As an illustration, let's say a client comes to you with the following request—*The traffic at the front of campus is too congested. I would like you to design a new traffic lane for northbound traffic exiting at the intersection at the front of the college.* So you design this new lane and have it added to the intersection. However, you find out three months later that the traffic congestion has decreased a little bit, but it is still a significant problem. So what went wrong? Clearly you did what was asked of you, but the problem was not solved, meaning that you were solving the wrong problem. The real problem was to improve the flow of traffic at the entrance. In this case, the client gave both the problem and the solution all in one statement. That is fine if a careful feasibility study was done and it was known that the additional traffic lane would alleviate the problem, but that was not the case here. This hypothetical situation is not so far fetched and happens in practice via neglect to do the up-front research or because underlying assumptions change. The point is that the <u>correct</u> problem should be identified and solved.

It would be better if the client had simply asked to improve the traffic flow, providing the opportunity to analyze the situation and develop different design options. Some questions to be asked in this situation are: *How much additional traffic is there? At what times does this happen? Where is the traffic coming from? What is an acceptable waiting time at the intersection?* It may be that several new lanes are needed, or perhaps the sequencing of the traffic signals is wrong, or maybe a new entrance could be added for less cost and improved traffic flow.

The lesson is that customers often come with the problems and solution all wrapped up together. When this is done, the *design space*, the space of all possible solutions to the problem, is unnecessarily limited. Be ready to tactfully challenge the assumptions and ask questions to get to the root of the problem. Ask clarifying questions, analyze, pick apart the request, and focus on the problem, not the solution.

Researchers and practitioners have examined the problem of eliciting needs, and it is an important pre-requisite for developing good engineering requirements specifications. Ulrich and Eppinger [Ulr03] proposed a process for obtaining the *voice of the customer* using the following five steps: 1) Gather raw data from users; 2) Interpret raw data in terms of needs; 3) Organize needs into a hierarchy; 4) Determine the relative importance of the needs; and 5) Review the outcomes and the process. Each of these steps is described in the following sections.

Step 1: Gather Raw Data from Users

This is often accomplished via interviews with supervisors, key users, or people from the client organization. In cases where new products are being developed, focus groups are often employed. The advantage of interviews and focus groups is that they provide the opportunity for dialogue with the user where new ideas, concepts, and needs may emerge. Another option is direct observation, where the team goes out and examines the system in use and develops concepts for improving it. IDEO Corporation is an innovative and successful company that designs new products and systems. They rely heavily on direct observation as a technique for successfully developing innovative products [Kel01]. For example, IDEO was asked by a client to develop a new medical instrument for balloon angioplasty used in hospital operating rooms. A critical requirement from the user was that only one hand could be used to operate the device because the technician's other hand had to be free during the procedure. From direct observation, the IDEO design team found that even though the current system was designed for one-hand use, it was impractical, and the technicians actually used both hands. IDEO designed and developed a two-handed pump that not only worked better than the one handed pump, but was quieter, easier to read, and had increased precision. This is another example of the customer specifying the solution as part of the problem statement.

Ulrich and Eppinger provide the following questions to ask during an interview:

- When and why do you use this type of product (system)?
- Walk us through a typical session using the product.
- What do you like about the existing products?
- What do you dislike about the existing products?
- What issues do you consider when purchasing the product?
- What improvements would you make to the product?

DILBERT® by Scott Adams

© UFS, Inc.

Figure 2.1 The difficulties of communicating with the customer. (Dilbert © United Feature Syndicate. Reprinted by permission.)

Step 2: Interpret the Raw Data in Terms of Needs

In this step the raw data is translated into customer needs. The needs are expressed in terms of what the system must do (a requirement) as opposed to how it is done. Statements of the customer's needs are known as *marketing requirements* or *marketing specifications*. For example, *"The system should have high-quality audio"* is a need or marketing requirement from the customer regarding performance, but says nothing about how it will be achieved. Marketing requirements are short sentences that describe the need in the language of the customer. They typically do not have a numerical target and are described as a state of being for the system. Other examples of marketing requirements are, *"The system should be easy-to-use,"* and *"The system should be able survive a drop from the runner's height."*

Step 3: Organize Needs into a Hierarchy

The marketing requirements are organized into a hierarchy of needs arranged from the most general to the most specific in successive levels of detail as required by the problem. It is organized by functional similarity, not as hierarchy of importance (that is the next step). This hierarchy is referred to as an *objective tree*. An example objective tree for a portable audio device intended for use by runners is shown in Figure 2.2. The three high-level objectives determined were high-quality audio, portable, and easy-to-use. Each of these is further sub-divided into the characteristics that support the higher level. For example, portability is divided into the needs of lightweight, small, ergonomic, and the ability to operate in the environment. The environmental need is further expanded into needs that support it.

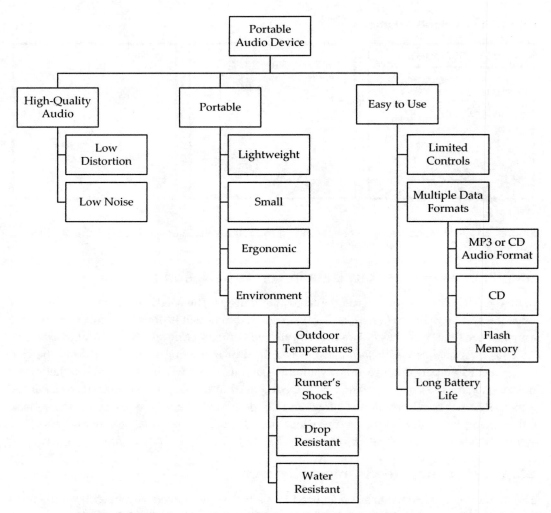

Figure 2.2 Objective tree for a portable audio device to be used by runners.

Step 4: Determine the Relative Importance of the Needs

The relative importance of the needs is determined based upon the experience of the designer and user needs. A simple method for ranking the needs, known as *pairwise comparison*, is presented. In pairwise comparison, all needs are systematically compared to all other needs at the same level in the hierarchy. A tool for facilitating the comparison process is the pairwise comparison matrix shown in Table 2.2. This matrix is created by first listing each of the needs as both row and column headings. Each cell in the matrix is evaluated by determining whether the associated row need is more important than the column need. If the row need is more important than the column, it is given a score of 1, if it is less important, it is given a score of 0. In cases where two needs are deemed to be of equal importance the entry is assigned a score of ½. Since it makes no sense to ask if a need is more important than itself, the major diagonal cells contain no entries. Once all of the cells entries are filled in, the overall score for each need is computed by summing all the values in its associated row. The results for the example presented in Table 2.2 indicate that portability is the most important need, while audio quality and ease-of-use are tied as the second most important. This is a very simple example, and most design problems will have more needs to compare. In addition, the needs at each sublevel in the hierarchy should be compared. The results of the ranking are used in later chapters to develop requirements specifications and compare design alternatives.

Table 2.2 Pairwise comparison matrix for ranking the highest-level needs of the portable audio device. This comparison should be carried out for all levels of the objective tree.

	High-Quality Audio	Portable	Easy-to-Use	Overall Score
High-Quality Audio	-	0	½	½
Portable	1	-	1	2
Easy-to-Use	½	0	-	½

Step 5: Review the Outcomes and the Process

The design process and all of its sub-processes are methods for making good decisions, and this technique for needs identification is no different. There is a certain amount of subjectivity and judgment that goes into it; the end result should be reviewed to determine if it makes sense. The objective is to challenge assumptions, fully identify the problem, and make informed decisions.

The three outcomes of this process are the marketing requirements that identify the needs, an objective tree that provides a hierarchical representation of the needs, and a ranking of the relative importance of needs. This process may seem as though it does not apply to student design projects, but in reality it does. The questions in this chapter are certainly candidates to ask when working on company-sponsored projects. If it is not a company sponsored project, the user needs should still be considered. For example, questions can be asked of friends and co-workers who are potential users of the system, focus groups can be formed, surveys administered, and Internet bulletin boards and discussion groups employed to gather this information.

2.4 Research Strategy and Resources

It is important to complete a thorough research survey while selecting and refining the project concept. Failure to do so may translate into time and money spent reinventing the wheel, while not taking full advantage of existing components, knowledge, practices, and technology. During the research phase, competing systems and technologies are identified, and based upon them the project concept refined, or in some cases abandoned. The character and strategy of the research survey is driven by the nature of the project. In general, the objective is to develop an understanding of the underlying scientific principles and demonstrate a familiarity with the state-of-the-art in the particular field. Some questions to be answered in the research survey are:

- What is the basic theory behind the concept?
- How is it currently being done?
- What are the limitations of current designs or technology?
- What are the similarities and differences between your concept and existing technologies?
- Are there existing or patented technologies that may be relevant to the design? If so, what are they and why are they relevant?

Internet Resources

The Internet is a powerful, fast, and readily accessible source for conducting research. There are many excellent search engines for locating web resources, but understand that it is important to go beyond the well-known search engines and beyond the Internet to conduct complete research.

One of the risks, and also one of the wonderful things, about the Internet is that virtually anybody can post information. It is important to analyze websites to ensure that they are reliable and credible. There are resources available that provide pointers on how to evaluate this credibility [Mci02, Sch98], and a little common sense goes a long way. One of the important things to look for is authorship—the author should be clearly identified and any affiliations listed. Carefully determine whether the information is subjective opinion or possibly a commercial for a product. Credible sites should provide references to original sources of material. Another step is to verify website content in print media or other reliable sources.

There are thousands of search engines available, making the task of selecting one challenging. Also, there are different types of search engines: text (search for the text or keywords; this can either be subject heading or full-page text search), indexed (the information is categorized into directories), meta-search (engines that search a variety of other engines), and natural language processing (allows questions with natural language queries). Many engines are some combination of the above and the following is a description of some of the available search engines. Be aware that this information may change rapidly and represents current information at the time of publication.

- AltaVista, www.altavista.com. This is a very complete resource that is good for focused research with specific search terms.

- AskJeeves, www.AskJeeves.com. Allows the user to ask questions using natural language queries, for example *"What is the difference between SRAM and DRAM?"*, *"Where can I buy Op Amps?"*, or *"Who won the 1950 World Series?"* It is good for questions that have probably been asked before, and all three of these queries produced useful information. Most importantly, the New York Yankees won the series by defeating the Philadelphia Phillies in four games, but it will also give you good references as to the differences between SRAM and DRAM.

- Google, www.google.com. Google is probably the most widely-used and well-known search engine. Google also has sub-search areas that are valuable for focusing a search such as www.google.com/unclesam that is targeted for finding US government information.

- Hotbot, www.hotbot.com. A good engine for full-page searches of specific search terms.

- Kartoo, www.kartoo.com. Unique in that it provides a graphical representation of the results and allows successive expansion of search subjects.

- Librarian's Index to the Internet, www.lii.org. A collection of Internet resources selected and evaluated by librarians, and according to their website, it is *a "well-organized point of access for reliable, trustworthy, librarian-selected Internet resources."*

- Yahoo, www.yahoo.com. Among the first and most well-known engines.

Electrical and Computer Engineering Resources

It is important to realize that the major search engines will not find all information available on the Internet. There are many websites with specialized search capabilities for design and electrical/computer engineering.

- The Chip Center, www.ChipCenter.com. A website for locating electronic components and their manufacturers. It provides links to product datasheets and application notes. It has a keyword search engine and a tree structure search for finding components. For example, you can start with Op Amps and delve into sub-categories such as Precision and High-Speed.

- Circuit Cellar, www.CircuitCellar.com. This companion website for Circuit Cellar magazine is a great reference for designers. It emphasizes embedded systems and electronics projects with many tutorial articles and project ideas.

- Datasheet Catalog, www.DatasheetCatalog.com. A datasheet source for electronic components and semiconductors.

- Dr. Dobbs, www.ddj.com. The magazine and companion website are a resource for software developers that includes tips and tutorials.

- EE Times, www.EETimes.com. Industry newspaper for electrical engineering field with information on current technology developments.

- Electronic Design Magazine, www.EDNmag.com. This free magazine is a resource for electrical design engineers that provides information on the latest manufacturer products. The website has a number of categorized technical resources and a useful design ideas section.

- ON Semiconductor, www.OnSemi.com. ON Semiconductor is a supplier of semiconductors for a wide range of applications, with a particular emphasis on power management issues. The website has a searchable database of over 15,000 components, and provides guidelines for component selection based on different technology applications.

- The Thomas Register, www.ThomasRegister.com. This is a source for finding companies and products in North America. It allows searches for parts and equipment that may be used in a design project. It provides profiles of companies that meet the search criteria and describes the products they make.

- In addition, most manufacturers of electronic components have websites providing product datasheets and application notes for their products. Application notes demonstrate how to use components in real applications. Some examples are Dallas Semiconductor, Fairchild Semiconductor, Motorola, and Texas Instruments.

Government Resources

- US Bureau of Labor Statistics, http://stats.bls.gov. This has valuable information on consumer spending information, allowing one to determine things such as how much people spend and what they spend it on. It also profiles specific industries and forecasts employment in different industry sectors.

- US Government Official WebPortal, www.FirstGov.gov. This is an entrance to all US government web resources.

- US Patent Office, www.uspto.gov. A searchable database of all patents back to 1790. Full text searches are available back to 1976 and full images back to 1790. It has information on the basics of patents, trademarks, and copyrights.

Journal and Conference Papers

A search will likely include journal and conference papers if technically detailed information on the latest theory or applications is needed.

- ACM (Association for Computing Machinery) Digital Library, www.acm.org. Provides abstracts (full text for subscribers) for ACM journals and conference proceedings.

- Compendex, www.engineeringvillage2.org. This provides indices to journal and conference papers in a broad scope of engineering fields, referencing material back to 1970.

- IEEE (Institute of Electrical and Electronics Engineers) Xplore Electronic Library, www.ieee.org. Provides abstracts (full text for subscribers) to all IEEE journals, transactions, magazines, and conference subscriptions published since 1988. In addition, abstracts for all IEEE standards are publicly available.

2.5 Project Application: The Problem Statement

A clear Problem Statement should be developed after the needs are identified, the initial research completed, and a project concept selected. The *Problem Statement* should identify the needs and the objectives of the project. The *needs statement* should briefly state the need being addressed and should not identify the solution. The *objectives statement* summarizes what is being proposed to meet the need, typically ranging from one or two sentences to one or two paragraphs in length. This can provide a preliminary description as to how the solution may be implemented, but a detailed description of the implementation should be avoided. The complete solution should not be posed until after the engineering requirements are fully determined. Three examples of needs and objectives statements, representing a cross-section of design projects, are provided in Examples 2.1–2.3.

Example 2.1 The Vision Enhanced Car Problem Statement. *Abstracted from the Vision Enhanced Car Design Report by Betts, Grayson, Haug, and Wykoff [Bet04].*

Need: Many drivers have difficulty seeing obstructions in all directions around their vehicle. The problem is exacerbated in larger sport-utility vehicles and trucks that have substantial blind spots. A system is needed to help drivers detect stationary objects, such as walls and curbs, and moving objects such as vehicles and people.

Objective: The objective is to design a system to enhance a driver's awareness of obstructions and help them avoid accidents. The system will detect objects around the sides and rear of a vehicle and alert the driver to hazards. It will access the vehicle's OBD II (onboard diagnostics) computer in order to monitor the vehicle's speed, and to determine whether it is moving in the forward or reverse direction. Thus, the system will provide alerts depending upon the operating conditions and the distance to detected objects. It will provide both audio and visual warnings when objects are detected. The user interface will be simple and unobtrusive so that it may be used while driving. The system should also be relatively easy to install. It should not interfere with the normal functioning of the vehicle and be as aesthetically pleasing as possible.

Example 2.2 Experimental Design Problem Statement. *Abstracted from the Intel Pro 1000XF Server Testing Design Report by Esek, Hunt, and Lewis. [Ese03].*

Need: Our customer is investigating the performance of commercial grade gigabit Ethernet fiber optic equipment for computer data communications in a military environment. The proposed system will utilize an Intel Pro1000 XF server card. This is a harsh operating environment and its effects on the performance and lifetime of the equipment are unknown. The client wishes to understand how the military environment affects the optical power margin of the Intel Pro 1000 XF card and associated connectors and cabling.

Objective: The goal of this project is to design the experimental equipment and test procedures to determine the effects of temperature variations and vibration on the optical power margin and the operating lifespan of the system.

Example 2.3 The RoboTender Design Problem Statement. *Abstracted from the Robo-Tender Design Report by Ciccozzi, Lamb, and Vilairoj [Cic05].*

Need: Many handicapped and elderly people find it difficult to prepare and obtain basic food and beverage items located in their home environment. This makes them dependent on friends, relatives, and in-home health care professionals to meet this basic need. With the current rise in health care costs and rise in the aging population, there is great potential for systems to help elderly and handicapped people meet this basic need.

Objective: The objective is to design and build a prototype system that will dispense basic beverages and food items, and by doing so achieve a level of independence for the target users. To meet their needs it should be affordable and operate on a kitchen countertop. It should also be easy to maintain and install. The system interface will be as simple as possible—it will be accomplished through a minimum number of push-buttons and/or voice recognition requests from the user. After the user successfully issues a request, it will dispense the item requested. The system will also provide clear visual and audio information on its status (food levels, errors, etc.).

A format for an *Extended Problem Statement* that integrates all of the elements of this chapter is as follows:

- *Need*. A statement that identifies the needs of the project.

- *Objective*. A statement describing the concept being proposed to meet the needs identified.

- *Background*. A summary of background research on the relevant technologies and systems. The objective here is to provide an introduction to the project and answer the questions posed in Section 2.4. The length and content of this section varies depending upon the project.

- *Marketing Requirements*. Short statements that describe the needs to be met in the language of the end-user.

- *Objective Tree*. A hierarchical representation of the needs based on functional similarity.

- *Ranking of the Needs*. The pairwise comparison matrices and the results of the scoring.

2.6 Summary and Further Reading

This chapter addressed the types of projects that are often undertaken by engineers and provides guidance in terms of questions that should be asked when selecting a project. The success of design projects depends upon adequately determining the user's needs and desires for the system. A process developed by Ulrich and Eppinger for needs elicitation was presented. The three outcomes of this process are: 1) marketing requirements that identify the customer needs, 2) an objective tree that provides a hierarchical representation of the needs, and 3) a ranking of the relative importance of the needs. It is important to conduct research on the concept and related technologies as the concept emerges and after the needs are identified. Pointers for conducting research were provided. Finally, a format for an extended problem statement was presented that summarizes the needs, objectives, and background research for a design project.

The works by Griffin and Hauser [Gri93] and Ulrich and Eppinger [Ulr03] are readable and more detailed discussions on how to obtain the voice of the customer. There are also other design books available that address how to identify needs and develop objective trees [Cro00]. Cagan and Vogel [Cag02] have proposed a process for product development known as iNPD (integrated new product development) and provide methods for navigating what they refer to as "the fuzzy front end of project definition."

2.7 Problems

2.1   In your own words, describe the differences between creative, variant, and routine designs.

2.2   List three guidelines that should be employed when selecting a project in practice.

2.3   Assume a customer comes to you with the following request—*Design a mechanical arm to pick apples from a tree*. What are the assumptions in this statement? Rewrite the request to eliminate the assumptions. (This problem was originally posed by Edward DeBono [Deb70]).

2.4   Assume a customer comes to you with the following request—*Design an RS-232 networked personal computer measurement system to transmit voltage measurements from a remote location to a central server*. What are the assumptions this statement? Develop a list of questions that you might ask the customer to further clarify the problem statement.

2.5   Describe what is meant by a marketing requirement.

2.6   What is the purpose of an objective tree and how is it developed?

2.7   The needs for a garage door opener have been determined to be: safety, speed, security, reliability, and noise. Create a pairwise comparison matrix to determine the scores and rankings of the five needs. Apply your own judgment in ranking the needs.

2.8   Consider the design of an everyday consumer device such as computer printer, digital camera, electric screwdriver, or electric toothbrush. Determine the customer needs for the device selected. The deliverables should be: 1) marketing requirements, 2) an objective tree, and 3) a ranking (pairwise comparison) of the customer needs.

2.9   **Project Application.** Select criteria to be applied for selecting a project concept as shown in Table 2.1 in Section 2.2. Candidate criteria are cost, time, and match with team member skills, to name a few. Then brainstorm to generate project concepts. Rank the top three to five concepts against the criteria selected as presented in Section 2.2.

2.10   **Project Application.** Determine the needs for the project selected. The result should be list of marketing requirements, an objective tree, and a ranking of the needs.

2.11   **Project Application.** Develop an Extended Problem Statement for your project concept as outlined in Section 2.6. Apply the processes presented in the chapter as appropriate for the project.

Chapter 3 The Requirements Specification

Specification • (n) A detailed and exact statement of particulars, a statement fully describing something to be built. —American Heritage Dictionary

The Requirements Specification identifies those requirements that the design must satisfy in order for it to be successful. It is in effect the mission statement that drives all subsequent stages of development, and when completed it should be a detailed and complete vision of the design goals. An effective Requirements Specification should identify all important requirements, yet provide enough flexibility for the design team to develop an innovative solution. It also serves as a communication tool for everyone involved in the design, such as engineering, marketing, and the client. All parties should agree to the content of the requirements before development proceeds. In some cases, the Requirements Specification serves as a legally binding agreement between the developers and the client.

A major challenge in developing the requirements is in many ways analogous to the proverbial *"What came first, the chicken or the egg?"* question. The final solution is analogous to the chicken, and the requirements are analogous to the egg. In the beginning, the chicken is hidden inside the egg, yet the egg must be capable of describing what the chicken will become. The difficulty is that it is hard to develop the *what* for the requirement without already having solved the problem or created the chicken.

This chapter guides the reader through the process of developing a Requirements Specification and is organized as follows. First, requirements terminology, which can be rather confusing, is defined, followed by a representation of the process for developing requirements. The properties of an engineering requirement are defined and examples for ECE systems are provided. This is followed by the properties of a complete Requirements Specification (the collection of marketing and engineering requirements), supplemented with examples and methods for developing and analyzing the requirements. The chapter concludes by addressing constraints and standards and their role in the Requirements Specification.

Learning Objectives

By the end of this chapter, the reader should:

- Understand the desired properties of an engineering requirement and know how to state them.

- Have a familiarity with engineering requirements that are commonly specified in ECE systems.

- Understand the desired properties of the complete Requirements Specification, as well as know how to develop and document it.

- Be able to analyze the tradeoffs between requirements.

- Be able to identify constraints and standards that apply to an engineering design problem.

- Be able to develop a complete Requirements Specification document.

3.1 Specification Terminology

Unfortunately, there is no standard set of terminology to describe the many different types of specifications. The objective of this section is to define some terms that are commonly used in practice and identify how they are used throughout this book

Marketing requirements, developed in Chapter 2, are statements that describe customer needs using language that the customer would use. Example marketing requirements for an audio amplifier are that *"the product should be lightweight"*, *"should have high-quality audio"*, or *"should be easy to use."* Marketing requirements are related to engineering requirements, and the link between them is examined in more detail later in this chapter. Alternate terms that are used are *raw requirements* (in IEEE standards) and *customer attributes characteristics* [Hau88], a term that is used in Quality Functional Deployment (QFD) methods that are addressed later in this chapter.

Engineering requirements are requirements that apply to the technical aspects of the design. Ideally, each engineering requirement is measurable and has an associated target value or range. They are also known as *engineering characteristics* in QFD. A *performance requirement* is an engineering requirement that is more limited in scope and specifies only performance-related measures. For example, performance requirements for a computer system would likely include measures of computational power and data access rates.

The *Requirements Specification*, which is the focus of this chapter, is a collection of engineering and marketing requirements that a system must satisfy in order for it to meet the needs of the customer or end-user. Alternate terms that are used for the Requirements Specification are the *Product Design Specification* [Pug90] and the *Systems Requirements Specification*

[IEEE Std. 1233-1998]. A Requirements Specification is the most general specification that applies to the entire product or system.

For complex systems, there are often multiple requirements for each of the constituent subsystems. Such specifications are known as *subsystem design specifications* or *component design specifications* [Pug90]. A *functional requirement* is an important subsystem design specification that is used in both the design and requirements phases. The functional requirement describes the required inputs, outputs, and functionality of a system or subcomponent. The functional requirement is employed in the development of the functional decomposition design method covered in Chapter 5.

Technical specifications are a complete list of technical details for a given system, such as operating voltages, processor architecture, and types of memory. This may include elements of the Requirements Specification, but reflects the state of final system realization. The technical specification is fundamentally different from a requirement in that it indicates what was achieved in the end versus what a system needs to achieve from the outset.

3.2 A Requirements Setting Process

Figure 3.1 illustrates the process for developing a Requirements Specification that was abstracted from the IEEE Guide for Developing System Requirements Specifications [IEEE Std. 1233-1998]. The three input sources are the customer, the environment, and the technical community. The input from the customer includes the marketing (raw) requirement, while the environment includes external constraints and standards, items that are addressed in this chapter. The input from the technical community is based upon the knowledge of the engineers. The heart of this representation is the central bubble for developing the Requirements Specification. This bubble is further expanded in Figure 3.2 and consists of identifying the engineering requirements, identifying the relationships between engineering and marketing requirements, and selection of quantifiable targets for each requirement.

Figure 3.1 Requirements Specification development processes abstracted from the IEEE Std. 1233-1998. The three input sources to the central process are the customer, environment, and technical community.

Figure 3.2 The process employed in this chapter to develop the system requirements.

3.3 Engineering Requirements

Before developing the complete Requirements Specification, we need to first address individual engineering requirements. This section identifies the desirable properties of engineering requirements and provides examples of them. This differs from the Requirements Specification, which is a collection of the individual requirements. An engineering requirement is a short statement that addresses a technical facet of the design. An example is *"The system should be able to supply 50W of power."*

3.3.1 Properties of an Engineering Requirement

Each engineering requirement should meet the four properties below [IEEE Std. 1233-1998]:

1) *Abstract.* This means that a given requirement should specify *what* the system will do, not *how* it will be implemented. This is the chicken and egg problem described earlier. It is frequently the most difficult property to satisfy since designers often have a preconceived concept for the solution. Unless absolutely necessary, the requirements should say nothing about the implementation. For example, a requirement stating that a certain microcontroller (i.e., technology) <u>will</u> be used should be avoided. Admittedly, this is not always possible due to customer requirements or in cases where a system is being built upon pre-existing technology. A common analogy used for the *"what versus how"* problem is that of designing a bridge. The requirement is to transport people from one side to other, without specifically stating the solution is a bridge, because another solution, like a ferry, may be a much more effective solution.

2) *Unambiguous.* Each requirement should have a single unambiguous meaning and be stated with short complete sentences.

3) *Traceable.* Requirements should be traceable to the wants and desires of the user. If the design doesn't satisfy the customer's needs, it won't be successful. Methods for documenting traceability are presented later in the chapter.

4) *Verifiable.* **Verifiability** means that there should be a way to measure or demonstrate that the target requirements are met in the final system realization. Doing so allows the system to be tested or verified against the requirements. The idea is that if there is no way to verify that the requirement is met, then it should not be a requirement. Verifiability is used to answer the question of *"Are we building the system correctly?"*

Let's examine an example requirement for a robot whose objective is to navigate autonomously within a specified environment. Consider the following requirement

> *The robot must have an average forward speed of 0.5 feet/sec, a top speed of at least one foot/sec, and the ability to accelerate from standstill to the average speed in under one second.*

Are the four properties for an engineering requirement met? In terms of the first property, abstractness, the answer is yes; it states what the system must do, not how it will be implemented. In terms of the second property, is it unambiguous? It gives clear bounds for speed and acceleration. The third property, traceability, can't be shown without the marketing requirements and is addressed later. Finally, can the requirement be verified? Speed and acceleration are directly testable in the final realization, and thus it is verifiable.

Now we analyze a second example requirement for the robot to see if it meets the properties

> *The robot must employ IR sensors to sense its external environment and navigate autonomously with a battery life of one hour.*

This requirement is not abstract since it identifies part of the solution in terms of the sensor type and the fact that batteries must be used. It is somewhat ambiguous in that it should specify what is meant in terms of autonomous and the operating period. In terms of operating period, should it work for exactly one hour and stop, or is greater than an hour acceptable? Again, traceability can't be demonstrated without the marketing requirements. This requirement would be hard to verify without a good definition of what autonomous navigation in this context means. A better requirement would be

> *The robot must navigate autonomously, with the aid of only landmarks in the specified environment, for a period of least one hour.*

In addition to meeting the four properties, requirements should be *realistic*. To be realistic, there should be a way of demonstrating that the target is technically feasible. Realistic targets can be determined with a little research, engineering know-how, creativity, or system modeling. One way to do this is to assume a solution for the final system (violating the abstractness property), for example, the weight of the robot, the motors used, the wheel size, and the battery selected impact the requirements. An engineering model based upon these characteristics could be developed to predict performance and support the realism of requirements. Alternatively, targets can be based upon prototyping, where a model or experimental system is developed to show that a particular requirement is feasible.

The use of benchmarking to identify similar systems and their performance provides a reference for realistic targets. It is generally hard to surpass the performance of well-developed products and systems on a first-generation design. An exception is with new and innovative approaches that allow you to surpass the competition. Competitive benchmarks may also be obtained from similar, but not necessarily identical, products. Experience working with a particular technology or previous generations of a system also provides guidance in selecting realistic targets. That being said, organizations wishing to gain or maintain a market edge often press the development team to achieve performance on new generations that were once believed to be unrealistic. Sometimes it just may not be feasible to verify the technical feasibility of requirements to determine if they can be met. In such cases, the requirements should have a certain amount of tolerance built into them and be updated as development proceeds.

3.3.2 Example Engineering Requirements

Many requirements may be specified for a design, but knowing which to include is the challenge. This section is meant as a guide to describe the types of engineering requirements that may be specified for ECE systems. **It is by no means definitive or inclusive of all possibilities, and the design team needs to determine those that are applicable to the particular situation. Attention must also be given to the verifiability of requirements for the particular application.**

Performance and Functionality

Performance requirements are critical measures that impact the acceptance of the product by the end-user, quantitatively describing the ability of a device to perform its intended function. They are among the most important requirements that need to be addressed. *Accuracy* specifies acceptable levels of allowable error. Examples are measurement accuracy, accuracy in part placement, or percent of time the system is correct. Bounds, or tolerances, are typically placed upon accuracy requirements. Example requirements are

> *The system should detect 90% of all human faces in an image.*

> *The system should be able to measure part location to within ± 1mm.*

Speed requirements answer questions such as how fast the system must operate, the rate at which it processes parts, or how quickly it responds. Examples are the processor speed of computers, the average access time for data retrieval, propagation delay for circuits, or the time in which a software system must respond to a query. Example requirements are

> *The system should retrieve the user data within three seconds.*

> *The system should be able to process video data at a rate of 30 frames per second.*

Of course, the specific performance requirements depend upon the system under consideration. For example, a common requirement for the performance of audio amplifiers is the following

The amplifier will have total harmonic distortion of less than 1%.

Functionality refers to functions that the system must carry out or things that it must accomplish. Examples are

The system will implement the left and right button functions of a standard mouse.

The system will convert ambient temperature to a digital readout with an accuracy of 1% over the measurement range.

Reliability and Availability

This refers to the expected period of time that a system will operate properly. Measures of *reliability* include failure rates and mean time to failure. Estimation of system reliability is given detailed coverage in Chapter 8. Example requirements are

The system will have a reliability of 95% in five years.

The system will have a mean time to failure of 10^6 hours.

This first requirement means that 95% of the systems should be properly operating (have not failed) in five years. The second describes the expected average time between failures.

Availability is closely related to reliability, but addresses the amount of time that a system is available for operation. Example availability requirements are

The system will be operational 99% of the time.

The system will be operational from 4AM to 10PM, 365 days a year.

Energy

Virtually all systems consume and/or produce energy and thus have energy requirements. *Energy consumption* refers to the amount of power that a system consumes, and may be specified in terms of maximum, minimum, or average values. Example requirements are

The system will have an average power consumption of 500mW.

The system will have a peak current draw of 1A.

Operating lifetime addresses how long the system will operate from a given power source. For battery-powered devices, operating time is critical, and the lifetime for a given source may be an important requirement. An example of such a requirement is

The system will operate for a minimum of three hours without needing to be recharged.

Source characteristics refer to the characteristics of the input and/or output sources, such as voltage, current, impedance, frequency, number of phases, and power requirements. An example requirement is

> *The system will operate from a 12V source that supplies a maximum current of 300mA.*

Environmental

This refers to the physical environment in which the system will operate. It is also reasonable to assume that this refers to the impact of the system on the environment, which is considered later in Section 3.6, Constraints.

Operating temperature range specifies the range of temperatures that the system must operate in. An example of this is

> *The system should be able to operate in the temperature range of 0°C to 75°C.*

Electronic component data sheets typically provide information on allowable operating temperature ranges, and can be used to determine the realism of the target or verify it in the final system. Depending upon the customer needs, the system may be tested in an environmental chamber to verify that the requirement is met. Humidity is similar in concept to the temperature requirement, addressing the required ambient humidity range. A system may also need to be water-resistant (withstand rain and snow) or waterproof (be submersible in water). An example requirement is

> *The system must be waterproof and operate while submersed in water.*

Depending upon the environment, the system may need to withstand *vibrations*. Bounds are typically specified in terms of frequency, magnitude, and duration of the vibration. An example requirement is

> *The system must be able to withstand vibrations of up to 60Hz with a peak magnitude of 1mm for a period of 1 minute.*

Electromagnetic Interference (EMI) results from any electromagnetic energy that interferes or disturbs the operation of an electronic system. Electronic systems may produce electromagnetic radiation and limits may need to be placed upon the amount of radiation emitted. EMI is typically measured with specialized testing apparatus. Conversely, a system may need to be able to operate properly given a certain level of EMI.

The system may need to withstand a specified amount of *shock* and still operate. This may be measured in G-force or via heuristics. An example requirement is

> *The system should withstand a drop from a height of six feet and still operate.*

Software and Computing

Software and computing requirements often apply to the system. Complete coverage of software requirements is beyond the scope of this book and is more formally addressed in the <u>IEEE Standard for Software Requirements Specifications</u> [IEEE Std. 830-1998] and the <u>Volere Requirements Specification Template</u> [Vol03]. Typically they are divided into *functional requirements,* that describe what the system must do, and *non-functional requirements* (or constraints), that indicate the impact of external factors on the system. Non-functional requirements are considered later in the chapter.

Requirements may be placed on data transfer rates, systems to which they interface, and data formats. An example is

> *The system will maintain a data transfer rate of 100Mbs.*

Usability requirements address the user interface to the software. Requirements can be placed on the number of menus in the system, an estimated learning time, and number and types of errors the user is allowed to make. An example requirement is that

> *Users of the system should be able to learn 80% of its functionality within two hours.*

3.4 Developing the Requirements Specification

The Requirements Specification is the complete set of all system requirements. This section identifies desirable properties of the Requirements Specification and provides three methods of increasing complexity for developing and documenting the requirements. Tools for analyzing tradeoffs between the requirements are also presented.

3.4.1 Properties of the Requirements Specification

The desirable properties of the Requirements Specification are as follows [IEEE Std.1233-1998]:

1) *Normalized (orthogonal) set.* There should be no overlap or redundancy between requirements. A mathematical analogy is that of orthogonal vectors. For example, the x and y axes of the two-dimensional Cartesian space are orthogonal vectors, meaning that the projection (dot product) of one vector onto the other is zero. Ideally, all requirements should be orthogonal with no redundancy.

2) *Complete set.* A complete Requirements Specification addresses all of the needs of the end-user and also those needs required for system implementation. Failure to define a complete set results in **under-specificity** where not all needs are met.

3) *Consistent.* The requirements should not be self-contradictory.

4) *Bounded.* The scope of the Requirements Specification should be identified. Determine the minimum acceptable bound for target values; going beyond what is necessary limits the design space of potential solutions. Applying unnecessary bounds results in **over-specificity**.

5) *Granular.* The level of granularity, or abstractness, at which to focus should be determined. If you are the system architect, then address system level needs, but if you are developing a subsystem or component, address the subsystem requirements that must be satisfied.

6) *Modifiable.* Requirements are typically considered to be evolutionary. This is because there are many unknowns at the start of a project, hence estimates for the requirements are made. The original requirements are known as ***baseline requirements***. The estimates can change as development proceeds, as long as the changes are communicated to and agreed upon by all affected parties. Versions of the requirements should be tracked and identified as modifications take place.

In addition to these properties, the following advice is provided. First, use the requirements document as a communication tool. Requirements should be developed and reviewed by all parties involved, including cross-functional units, such as engineering, marketing, and manufacturing. Second, document the justification for selecting the requirements. Requirements are often determined by a team based upon many discussions. Documenting the justification is valuable for future reference when changes to the requirements may need to be made. Finally, realize that experience is an important part of the process and practice improves the ability to develop requirements. Most novice designers struggle, but perseverance and practice pays off.

A term that was introduced earlier is verification, which seeks to answer the question of whether or not the system is being developed correctly. Verifiability was applied to individual engineering requirements. A related concept is that of validation, which seeks to answer the question *"Are we building the right product?"* More formally, ***validation*** is the process of determining whether the system meets the needs of the user—is it valid? This is more general in scope than verification and more difficult to show. Requirements validation is usually carried out by reviews of the requirements by a team of people. Validation is demonstrated by being able to answer the following questions in the affirmative [Som01]:

- For each individual engineering requirement, are the traceability and verifiability properties met? Is each requirement realistic and technically feasible?
- For the Requirements Specification, are the properties of orthogonality, completeness, and consistency met?

A complete Requirements Specification includes all the requirements, both marketing and engineering, along with the relationships between them. The relationships between the engineering and marketing requirements need to be described to ensure that all the marketing specifications are being addressed by design. The relationship between the marketing and engineering requirements is called a mapping because, like a mathematical mapping, it defines which elements of the domain (marketing specifications) are associated with which elements of the range (engineering specifications). Three different mappings are presented that can be applied to meet the needs of a given problem.

3.4.2 A 1-to-1 Mapping

An example Requirements Specification for an audio power amplifier is shown in Table 3.1. The table has columns for the marketing requirements, engineering requirements, and for the justification of the engineering targets. The marketing requirements are short statements describing the end-user needs as presented in Chapter 2. The engineering requirements provide quantifiable targets and each should follow the properties identified in Section 3.3. The justification is used to indicate why the requirement is realistic. In this case there is a 1-to-1 mapping between marketing and engineering requirements. The term mapping is used to denote relationships between the requirements. If changing one requirement affects the other, then they are mapped, while requirements that are independent of each other are not mapped. A 1-to-1 mapping means that each marketing requirement has one, and only one, associated engineering requirement, and vice-versa.

Table 3.1 Requirements for an automobile audio power amplifier that employ a 1-to-1 mapping. The engineering targets are indicated in italics. The justification indicates how the target value for the engineering requirement was selected.

Marketing Requirement	Engineering Requirement	Justification
Should have excellent sound quality.	*Total harmonic distortion* (THD) should be ≤ 1%.	Based upon competitive benchmarking and existing amplifier technology. Class A, B, and AB amplifiers are able to obtain this level of THD.
Should have high output power.	Should be able to sustain an *output power* that averages ≥ 35 watts with a peak value of ≥ 70 watts.	This power range provides more than adequate sound throughout the automobile compartment and is competitive in this price range.
Should be easy to install.	*Average installation time* for the power and audio connections should not exceed 5 minutes.	Past trials using standard audio and power jacks demonstrate that this is a reasonable installation time.
Should have a reasonable cost.	*Production cost* should not exceed $100.	This is based upon competitive market analysis and previous system designs.

These requirements are fairly simple, with four marketing requirements and engineering requirements. Audio power amplifiers are widely available devices, so the requirements were determined through competitive benchmarks and knowledge of amplifier circuit designs.

The first engineering requirement directly impacts sound quality and is known as *total harmonic distortion* (THD). THD measures how closely the amplifier output signal follows the input signal. It is desirable for an amplifier to have a linear relation between input and output, where the output signal is identical to the input signal, except for an amplification factor. In reality, all amplifiers have some degree of nonlinearity or distortion. It is measured by applying a pure sinusoid as the input to the amplifier, which in the case of a perfectly linear amplifier produces a pure output sinusoid of the same frequency. Any nonlinearity introduces unwanted harmonic frequencies. THD represents the power of unwanted harmonics relative to the power of the fundamental sinusoid. THD is typically less than 1% for a good amplifier.

The second requirement, *output power*, is quantified in terms of both average and maximum values to minimize ambiguity. The third marketing requirement, ease of installation, is perhaps easy to understand intuitively, and is quantified by an engineering requirement for the expected installation time. The final requirement, cost, was selected based upon competitive benchmarks and existing designs.

3.4.3 A 1-to-N Mapping

A 1-to-1 mapping between engineering and marketing requirements is often unrealistic because each marketing requirement is unlikely to have a unique mapping to a single engineering requirement. For example, sound quality in the amplifier example is impacted by more engineering requirements than THD. As such, multiple engineering requirements may impact a single marketing requirement, producing a 1-to-N mapping between marketing and engineering requirements.

Table 3.2 contains example requirements developed for the design of an experimental test setup (the audio amplifier is revisited in the next section). The problem statement for this example was presented in Chapter 2 (Example 2.2), where the objective was to design a system to test a gigabit Ethernet card for use in a harsh operating environment. In particular, the effects of temperature and vibration variations on the optical power margin, both of which impact the bit-error rate and the system performance, were determined. The engineering requirements were selected based upon characteristics of the operating environment, through discussion with the client (in this case engineers), and via some educated guesswork.

Let's consider some of the requirements, starting with the effects of temperature variation on the optical power output. The testing requirement on the temperature range was selected based on the operating environment, while the accuracy is driven by the test equipment. The requirements also address the vibration testing requirements, including the vibration frequency range, amplitude, and resonant frequency. Note that the requirements are abstract—they describe what the system must achieve, not how it will be designed.

Table 3.2 Requirements for an experimental design project that employ a 1-to-N mapping. The objective was to design a system to test a gigabit Ethernet card for use in a harsh operating environment. The example is a modified version of the requirements presented in [Ese03].

Marketing Requirement	Engineering Requirement	Justification
The measurement of optical power should be accurate.	Must be able to measure the *optical power output* with an *accuracy* of ± 0.5dB.	This is based upon commercially available optical power measurement instruments.
Measure the effects of temperature variations on the optical power output.	Must be able to measure the *optical power output* from 10°C to 55°C.	This range simulates the operating environment, and 55°C is the maximum operating temperature of the card.
	The system must maintain *temperature accuracy* to within ± 1°C during all tests.	Based upon accuracy of commercially available test chambers.
Measure effects of vibration on the fiber optic connector and optical power output.	Must be able to measure optical power over a *frequency range* from 4Hz to 33Hz in increments of 1Hz.	The frequencies encountered in actual operation will not exceed this range.
	The *peak vibration amplitude* should be 0.01 inches.	The amplitude in the operating environment will not exceed this value.
	The card should be tested at a given frequency for a *duration* of 1 minute.	This exceeds the expected duration of vibration at given frequency that the system will encounter.
	The vibration effects should be tested in x, y, and z *directions*.	The system will encounter vibrations in multiple directions. This will provide data on differences in directional variation due to vibration.
	The experiment should determine *resonant frequency* to an accuracy of ± 0.5Hz.	This will provide data on worst case vibration at the resonant frequency.

3.4.4 An M-to-N Mapping

The most general case is an M-to-N mapping, where any marketing requirement can be mapped to any engineering requirement and vice versa, allowing complete flexibility. This is a realistic scenario for many systems and is a straightforward extension of the previous cases. Example requirements for the audio power amplifier are shown in Table 3.3. The marketing requirements are the same as before, but the engineering requirements are expanded to be more realistic by including power efficiency (ratio of power consumed to power delivered to the speakers) and the amplifier dimensions. The M-to-N mapping is achieved since any engineering requirement can be mapped in the table to any marketing requirement and vice versa.

Table 3.3 Requirements for the audio power amplifier that employ an M-to-N mapping. The marketing requirements are listed at the end of the table.

Marketing Requirements	Engineering Requirements	Justification
1, 2, 4	The *total harmonic distortion* should be <0.1%.	Based upon competitive benchmarking and existing amplifier technology. Class A, B, and AB amplifiers are able to obtain this level of level of THD.
1–4	Should be able to sustain an *output power* that averages ≥ 35 watts with a peak value of ≥ 70 watts.	This power range provides more than adequate sound throughout the automobile compartment, and is competitive in this price range. It is a sustainable output power for projected amplifier complexity.
2, 4	Should have an *efficiency (η)* >40 %.	Achievable with several different classes of power amplifiers.
3	*Average installation time* for the power and audio connections should not exceed 5 minutes.	Past trials using standard audio and power jacks demonstrate that this is a reasonable installation time.
1–4	The *dimensions* should not exceed 6" x 8"x 3".	Fits under a typical car seat. Prior models and estimates show that all components should fit within this package size.
1–4	*Production cost* should not exceed $100.	This is based upon competitive market analysis and previous system designs.

Marketing Requirements
1. The system should have excellent sound quality.
2. The system should have high output power.
3. The system should be easy to install.
4. The system should have low cost.

Let's examine the first engineering requirement for THD, which is the same requirement that was in Table 3.1. Here it is mapped to the marketing requirements for sound quality, output power, and low cost. It is clear that THD is directly related to the sound quality. However, it is also impacted by the output power, since high output power can be achieved at the expense of THD. In addition, better THD can be achieved through the use of more and better quality components. The tradeoffs between the engineering and marketing requirements can be quite complex and are examined more fully in the following section.

3.5 Advanced Requirements Analysis

This section examines tools that are used to analyze and refine the requirements, and can be applied to any of the mappings presented. As we saw in the example in Table 3.3, there are many tradeoffs between the different system requirements. Understanding the tradeoffs is valuable for developing a better understanding of the system and for refining the requirements themselves. In addition, the benchmarking of requirements is addressed.

3.5.1 The Engineering-Marketing Tradeoff Matrix

This matrix identifies how engineering and marketing requirements impact each other. To demonstrate its construction, we continue to examine the automobile audio amplifier example. The tradeoff matrix is shown in Table 3.4, where the marketing requirements constitute the row headings and the engineering requirements the column headings.

Table 3.4 Engineering-marketing tradeoff matrix for the audio amplifier. ↑=positive correlation, ↑↑=strong positive correlation, ↓=negative correlation, ↓↓=strong negative correlation.

Marketing Requirements	Polarity	THD (−)	Output Power (+)	η, Efficiency (+)	Install Time (−)	Dimensions (−)	Cost (−)
1) Sound Quality	+	↑↑	↓			↓↓	↓↓
2) High Power	+	↓	↑↑	↑		↓↓	↓
3) Install Ease	+		↓		↑↑	↑	↓
4) Cost	−	↓↓	↓	↓		↓	↑↑

One of the first things to note is that each requirement has an associated polarity. A requirement with a positive/negative polarity, denoted with a +/- symbol, means that increasing/decreasing that requirement increases the desirability of the product, respectively. A goal is considered a requirement with its polarity. For example, cost almost universally has a negative polarity because decreasing cost almost always makes a product more desirable.

The entries in the body of the matrix can be thought of as a correlation that measures the ability to achieve the marketing and engineering goals simultaneously. A positive correlation

(↑) means that both goals can be simultaneously improved, while a negative correlation (↓) means that improving one will compromise the other. Not all correlations are of equal importance, the strength of the correlation being denoted by the number of arrows. Blanks entries in the matrix mean that there is no correlation between the requirements.

To better understand the matrix, consider the entries in the top row associated with sound quality. The relationship between THD and sound quality is denoted by the double positive arrow. This relationship is interpreted as

> *The goal is to increase sound quality and decrease THD. There is a strong positive correlation between them since decreasing THD increases sound quality.*

What wasn't so clear in the 1-to-1 mapping is that there is a link between the goal of maximizing output power and the goal of maximizing sound quality (second entry in the top row). This entry is interpreted as

> *The goal is to increase sound quality, to increase output power, and there is negative correlation between them since increasing output power will decrease sound quality.*

That is because electronics can be designed to achieve larger output power at the expense of sound quality. However, it gets a little more complicated, since output power can be increased without loss in sound quality, if more amplifier stages are employed. That increases the dimensions and cost, thus the relationships between sound quality, cost, and dimensions are identified in the first row. When creating the entries in the matrix, it should be assumed that only the associated requirements can vary and that all others are held constant. When finished, the matrix allows a quick and easy reading of the tradeoffs between engineering and marketing requirements. This example demonstrates the complex nature of a seemingly simple device, and provides a much clearer picture of the design tradeoffs involved.

3.5.2 The Engineering Tradeoff Matrix

The example of output power in the previous section illustrates the need to examine the tradeoffs between the engineering requirements, which are shown in Table 3.5. In this table the engineering requirements constitute the headings for both the row and column entries. Only the entries above the upper diagonal elements are filled in due to the redundancy of the lower diagonal elements. Again, positive and negative correlations are indicated along with the strength of correlation.

Let's examine the tradeoffs involved with output power. High output power can be achieved at the expense of THD as shown in the first row of the table. The second row indicates that there is a positive correlation between efficiency and output power, since the more efficient an amplifier is, the more power it can deliver. There is a negative correlation between the dimensions and power, since larger parts and greater surface area aid in dissipating more power. Finally, there is a negative correlation between the output power and cost because of the greater size and number of parts needed to achieve higher power.

Table 3.5 The engineering tradeoff matrix for the audio amplifier. ↑=positive correlation, ↓=negative correlation.

		THD	Output Power	η, Efficiency	Install Time	Dimensions	Cost
		-	+	+	-	-	-
THD	-		↓			↓	↓
Output Power	+			↑		↓	↓
η, Efficiency	+					↑	↓
Install Time	-					↓	
Dimensions	-						↓
Cost	-						

3.5.3 Competitive Benchmarks

Competitive benchmarking helps to select targets for the engineering requirements. By analyzing competing systems, a better understanding is gained of what is realistic and where the design may potentially outperform the competition. The benchmark table lists the requirements in the row headings and the competitors in the column headings as shown in Table 3.6.

Table 3.6 Competitive benchmarks for the audio amplifier.

	Apex Audio	Monster Amps	Our Design
THD	0.05%	0.15%	0.1%
Power	30W	50W	35W
Efficiency	70%	30%	40%
Cost	$250	$120	$100

3.5.4 The House of Quality

A well-known tool for developing requirements is the House of Quality (HOQ). The HOQ is part of a product development process known as Quality Functional Deployment (QFD) that is widely used in industry. QFD is a series of processes for product development that incorporate the needs of the customer throughout the system lifecycle. It encompasses design, manufacturing, sales, and marketing. QFD is characterized by a series of matrices that have a visual appearance similar to that of a house. The matrices relate different aspects of the development process and are effective for communicating between different units in an organization. There are houses for different phases of product development, but here the focus is on using the HOQ for the Requirements Specification. A HOQ for the audio amplifier example is shown in Figure 3.3. It contains all of the elements that we have addressed so far—marketing requirements, engineering requirements, engineering-marketing tradeoffs, engineering tradeoffs, and the target values for the engineering requirements. The HOQ is presented for completeness, but is redundant since it contains all of the information already presented in Tables 3.3–3.5. The HOQ also becomes visually overwhelming and hard to read as problem complexity grows.

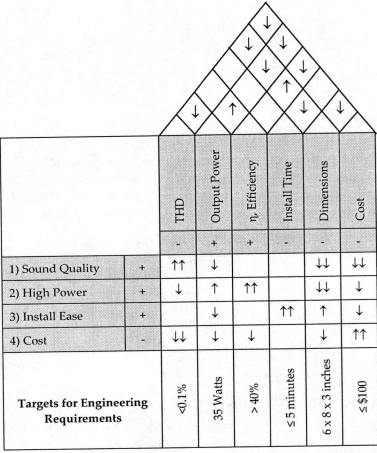

		THD	Output Power	η, Efficiency	Install Time	Dimensions	Cost
		-	+	+	-	-	-
1) Sound Quality	+	↑↑	↓			↓↓	↓↓
2) High Power	+	↓	↑	↑↑		↓↓	↓
3) Install Ease	+		↓		↑↑	↑	↓
4) Cost	-	↓↓	↓	↓		↓	↑↑
Targets for Engineering Requirements		<0.1%	35 Watts	> 40%	≤ 5 minutes	6 x 8 x 3 inches	≤ $100

Figure 3.3 The complete House of Quality for the audio amplifier example. This integrates the information in Tables 3.3, 3.4, and 3.5.

3.6 Constraints

One of the drivers of the requirements development process that we saw earlier in Figure 3.1 was the operating environment. The operating environment contributes constraints to the Requirements Specification. A *constraint* is a restriction that impacts the design but is not a specific function that the system must deliver. They are sometimes referred to as *non-functional requirements*. Constraints and engineering requirements are closely related and may be the same, but constraints are treated as a separate entity for two reasons. First, constraints are broader in scope than engineering requirements, including social, ethical, and political considerations that may not be verifiable. Second, constraints ensure that a problem is not solved in a vacuum, but in the context of society, industry, and in a sustainable ethical fashion.

The ABET accreditation agency for engineering programs requires that all capstone projects incorporate constraints for exactly that reason. ABET recommends that projects address most of the following constraints: economic, environmental, ethical, health and safety, manufacturability, political, social, and sustainability. This section provides guidance on the meaning of the constraints in an ECE context and how to address them. Constraints should be clearly identified using short statements similar to those used for engineering requirements. **The examples provided are by no means definitive or inclusive of all possibilities, and the design team needs to determine those that are applicable to the particular situation.**

Economic

Economic constraints include the costs associated with the development (design, production, maintenance) and sale of a system. They may also include the economic impact of the final system, such as how they are going to contribute to profits or save the user money. Example economic constraints are

> *The costs for developing the system should not exceed $50,000.*
>
> *The final sales cost should be less than $1,000.*

Environmental

This refers to the environment in which the system will operate, and includes temperature range, humidity, vibration, electromagnetic interference immunity, and shock. Quantitative environmental requirements were provided in Section 3.3. Other examples are

> *The system should be able to function in high dust and high humidity environments.*
>
> *The system should be able to function properly in the automotive passenger cabin.*

Another perspective is in terms of the impact of the design on the external environment and the wise usage of the earth's resources. For example, power efficiency is the percentage of input power that is effectively used by the system, the remaining amount which is lost and generally to be minimized. This addresses sustainability in terms of sustaining and conserving energy resources. An example is

The system will use 35% less energy than competing products and qualify for US Energy Star certification.

Recyclability is the ability to dismantle a product into its constituent materials for reuse in other products. European countries have regulations on the recyclability of consumer products. In many cases, the producer of a product is responsible for its safe disposal once its service life is over.

Ethical and Legal

Designs should not infringe upon existing patents, copyrights, and trademarks, particularly if the intention is to sell the product. Patent searches should be conducted, and search capabilities are available at the United States Patent Office website (www.uspto.gov). An example constraint is

An intellectual property search will be conducted to ensure that there is no infringement on prior patents.

Security and privacy constraints apply to systems that handle sensitive data or personal communications. The ability of computing systems to withstand malicious attacks by hackers is another consideration, and the use of firewalls or other protective measures may be warranted. Examples are

The system will protect the user's identity with 128-bit encryption.

Only managers will be allowed access to personnel data.

When modifying an existing system, care must be taken so that changes do not adversely affect the performance or operation of the system. Also, the design should not adversely interfere with the operation of other systems. An example is

The system should not interfere with the communications of other mobile devices.

Health and Safety

The health and safety of anyone affected by the final product is an especially important consideration. An obvious electrical hazard is related to high voltage and high current situations. An example constraint is

The system will minimize the user's exposure to electrical shock.

In the design phase, this can be addressed through the use of ground fault protection, fuses, breakers, and overload protection circuitry. Strong electrical fields also pose risks, and exposure levels need to be considered. For example, IEEE and ANSI standards provide guidance on safe levels for exposure to radio frequency electric fields.

There is a tendency to think that physical harm is not an issue in ECE systems, but many electronic systems control mechanically moving parts. Consider the design of an automatic garage door system. An example constraint could be that

The system should bring no harm to people or objects in the door's path.

This could translate into engineering requirements on the ability to automatically detect objects in the door path and the ability to stop the door from closing if it exerts a certain amount of force on an object or person.

There are many safety standards, and two that are widely applied for consumer products are the UL (Underwriters Laboratory) and CE (Common European) standards. Example constraints are

The system will use only UL approved components.

The final system will be tested at an independent laboratory for UL and CE approval.

Manufacturability

A prevalent product development paradigm that used to be employed in many engineering organizations was to "*throw the design over the wall.*" What this meant is that the design and development team would create a new product and hand it off to the manufacturing team to produce (throw it over the wall and run), often without having considered the manufacturability of the product. The manufacturing team would then address how to produce the design, and in many cases could not do so without major redesign. Fortunately, this has given way to much better concurrent engineering practices where all aspects of product development are considered throughout the process.

Size is a consideration in terms of the amount of space the final design will occupy, particularly if it has to be physically integrated with other components. An example constraint is

The system must be manufactured on a circuit board with dimensions of no greater than 1"x2".

The realization and portability of the system is a consideration. For example, with electronics, will it be built on a printed circuit board? Following design rules and file format guidelines is important for manufacturing printed circuit boards and integrated circuits. A chip foundry may require that integrated circuit layouts utilize certain file formats. Examples of realization constraints are

The system will be manufactured using three layer printed circuit board technology.

The software will be distributed via the Internet.

The product should be run on both Linux and Microsoft operating systems.

The use of readily available parts, instead of low volume or hard to find components, improves manufacturability, and an example is

The design shall only incorporate components that can be purchased through two of our main suppliers.

Political

Political constraints address relationships to political, governmental, or union organizations. Examples include obtaining governmental approvals, resolving trade barriers, and determining the acceptance of systems for use in unionized environments. Examples are

The system will need to obtain FDA approval before it can be sold to medical users.

The software will comply with the Digital Millennium Copyright Act.

Social

This addresses aspects such as benefits, risks, and acceptance of products by the intended user or by society at large. For example, robots have tremendous benefits for improving product quality, while freeing people from dangerous and repetitive tasks. Yet when used in automation, they present the risk of displacing workers and causing job losses.

Many great products have fallen by the wayside because users were unwilling to accept it. An example is the early Apple Newton Personal Digital Assistant, the first product of its kind. The fatal flaw was handwriting recognition that required a training process for accurate recognition. This was not accepted by consumers and Palm® Computing solved the problem by employing a simplified alphabet known as Graffiti. Graffiti was also seen as risky when it was being developed, but due to its simplicity it was accepted by consumers and the product became a huge success. In the mid 1980s, Phillips electronics released the Laser Disk player, which failed magnificently, but was far superior to VHS technology. Their failure was attributed to the cost of the players and disks when compared to VHS. Fifteen years later DVDs, with the help of computers, reached a price point which now makes them preferable to VHS.

Will the system be used by engineers, technicians, laborers, doctors, lawyers, or the general public? Each group has its own culture, educational background, and willingness to accept innovations. Example constraints are

The product shall provide help menus to the user in either English or Spanish.

The software will be designed to easily be used by operators on the manufacturing floor.

Sustainability

Sustainability refers to the sustainability and maintenance of the system being developed, or how the product itself will be sustained. Lifetime is an estimate of the operating lifetime of the device or system, and an example is

The expected operating lifetime of the system is 10 years.

Will the system be designed so that it can be reused in future applications? This is common in software development where the objective is to design modules that are reliable and flexible enough to be used in other applications. It is also a consideration in terms of the reusability of electronic or digital components in future system upgrades. An example reuse constraint is

> *The software should maintain downward capability and be able to use version 2 object libraries.*

After a product goes into service, it enters the maintenance phase, where it is maintained and upgraded. In software designs this is an important consideration, as software is regularly upgraded and maintained. On the hardware side, maintenance can be facilitated by the use of plug-in modules that are easily removed and replaced. Example maintainability constraints are

> *The system will initially be available to 100 users at five field locations, and within one year must be expanded to address usage by 5,000 users company-wide.*

> *The system should have a modular design such that failed components can be replaced by a technician in under 15 minutes.*

There may be internal restrictions on system development imposed by the company based on their internal expertise and ability to maintain the system, such as

> *The system will use only PIC microcontrollers.*

3.7 Standards

Standards are exactly what the name implies, a standard or established way of doing things. Without standards, the use of technology would be severely limited, if not downright impossible. Standards ensure that products work together, from home plumbing fixtures to the modules in a modern computer. Imagine if every computer manufacturer had their own communication standard, instead of following established protocols such as RS-232, TCP/IP, and USB—your computer would have a hard time printing, sending email, instant messaging, or surfing the Internet! Furthermore, standards ensure the health and safety of products that people use every day. Identifying and following standards is an expected part of good engineering practice.

3.7.1 How to Use Standards

The question becomes, what standards are relevant to your project and how do you use them? There are different levels of interaction with standards that we denote as: user, implementation, and development levels. At the *user level*, the standards are simply employed in the design, and detailed technical knowledge of the standard is typically not necessary. For example, when using a component that communicates to other devices, it is likely that a standard communication protocol is used. Other than having to configure software or hardware to communicate with the standard, detailed knowledge of the standard isn't required. Another example would be in developing software to display digital images in a standardized format such as JPEG-2000 (Joint Photo Experts Group), in which case it is likely that existing software components would be used to read and display data in this format.

At the *implementation level* details of the standard need to be understood. For example, when developing low-level drivers for computer peripherals, you need to become an expert on the underlying standard. Another example is reliability, where the requirement may be that *"the system will have a reliability of 95% in 10 years."* In this case a reliability standard, such as <u>Military Handbook for Reliability Prediction of Equipment</u> [MIL-HDBK 217F] may be employed, and its usage requires an understanding of both the reliability theory and the standard itself.

New standards are constantly being developed and existing ones modified, leading to the final level of interaction at the *development level*. Depending upon the standard, engineers from different organizations, professional societies, and corporations take part in the standards setting process. Many participants in this process are trying to gain a competitive advantage for their products and services.

It is difficult to navigate the world of standards; they tend to be highly detailed and limited parts of a standard may apply to a project. In addition, many standards are costly to obtain, while some are freely distributed. The following is some advice for identifying and employing standards. First, conduct research on applicable standards. Virtually all standards organizations maintain websites that provide basic information on their particular standards. The IEEE Xplore database is a good place to start since it has a wide variety of standards and provides free searchable abstracts. Many companies and universities have subscriptions databases of complete standards. Second, determine the expected level of interaction. Based upon your analysis of the problem, do you foresee applying standards? Or will you need to develop an in-depth knowledge at the implementation level? In the latter case, you need to obtain detailed information on the applicable standards. Finally, you should consider asking your client. They may have their own internal standards and procedures to follow, and they may have experts on the applicable standards.

3.7.2 Types of Standards

This section provides an overview of different types of standards that may be employed in a project.

- *Safety*. Safety standards address how to design for safety, and how to test products to ensure that they are safe. As indicated in the constraints section, this may be in terms of electrical shock hazards, and exposure to electromagnetic radiation.

- *Testing*. Testing standards are often related to safety, but are broader in scope. For example, standardized benchmark tests are used for comparing computational performance, one well-known standard being the SPEC (Standard Performance Evaluation Corporation) benchmarks.

- *Reliability*. Reliability standards address general reliability principles and design methods for different classes of systems. Another practical aspect is in the estimation of reliability of electronic systems, such as the IEEE and military reliability standards.

- *Communications*. They address how electronic systems communicate and transfer information, such as in computing, telephony, and satellite communications.

- *Data Formats*. Standard data formats ensure that systems and software can properly share information. Examples include image, video, and database standards.

- *Documentation*. There are standards for technical report documentation. In addition, there are standards for documenting processes and business practices, a well-known case being the ISO (International Standards Organization) 9000 and subsequent standards.

- *Design Methods*. Certain design techniques are standardized as well. Examples include software design methodologies, and the use of design languages such as the Hardware Description Language (HDL) and the Unified Modeling Language (UML).

- *Programming Languages*. Programming language syntax is standardized so that software maintains a level of portability between systems and compilers.

- *Connector Standards*. Standards for cable connections are common and should be followed to ensure that systems are easily interfaced and manufactured.

- *Meta-Standards*. Some standards are a combination of multiple standards known as meta-standards. For example, the RS-232 standard is really a combination of a mechanical standard describing the connector physical dimensions connector, an electrical standard describing the voltages, a functional standard describing the pins and their function, and a procedural standard describing how entities communicate.

3.8 Project Application: The Requirements Specification

The following is a recommended format for a Requirements Specification document that also integrates the problem statement from Chapter 2.

- *Needs, Objectives, and Background*. Include the elements defined in the Problem Statement in Chapter 2.

- *Requirements*. Identify the marketing requirements, engineering requirements, and justification in a table format (see Tables 3.1, 3.2, and 3.3). The particular method selected depends upon the complexity and needs of the problem. Supplement this with tradeoff matrices (Tables 3.4 and 3.5) and competitive benchmarks if necessary.

- *Constraints.* Identify applicable constraints from the categories of economic, environmental, ethical, health and safety, manufacturability, political, social, and sustainability. Some of the constraints may overlap with the engineering requirements. Remember that for ABET accredited programs, capstone projects should address most of the constraints identified in Section 3.6.
- *Standards.* Identify standards that may be applicable to the project and how they will impact the design.

Table 3.7 presents a self-assessment checklist for the Requirements Specification.

Table 3.7 Self-assessment checklist for the Requirements Specification. 1 = Strongly Disagree, 2 = Disagree, 3 = Neutral, 4 = Agree, 5 = Strongly Agree.

Engineering Requirements	Score
Each engineering requirement is abstract and does not provide the solution.	
Each engineering requirement is unambiguous and is written as a concise statement.	
Each engineering requirement can be traced to a user need.	
Each engineering requirement is verifiable.	
Each engineering requirement is realistic and has a justification provided.	
The Requirements Specification	
The requirements are normalized, with minimal redundancy and overlap.	
The requirements are complete, addressing the needs of the user and providing guidance for system implementation.	
The requirements are bounded (not over-specified).	
The tradeoffs between requirements (marketing-engineering) and (engineering-engineering) have been addressed.	
Constraints and Standards	
The constraints that apply to the project are clearly defined and relevant to the problem.	
All of the eight constraint categories have been assessed relative to the design.	
The constraints are stated unambiguously with concise statements.	
Standards that may be applicable to the project have been researched and identified.	

3.9 Summary and Further Reading

This chapter presented a process for developing the Requirements Specification, which consists of iteratively identifying the requirements, mapping them to the customer needs or marketing requirements, and selecting the targets as indicated in Figure 3.2. The desirable properties of engineering requirements and the complete Requirements Specification were presented. The verification of a requirement is particularly important, as it seeks to help in answering if the system is being built correctly. Requirements validation addresses whether the requirements meet the needs of the user, or if the correct product is being designed. Methods of documenting the requirements include 1-to-1, 1-to-N, and M-to-N mappings of engineering and marketing requirements. Tools for benchmarking and analyzing the tradeoffs between requirements were given. Proper determination of the requirements significantly influences all subsequent phases of the design, thus the final requirements document should be agreed upon by all stakeholders.

The processes presented here were developed from research in the field and the authors' teaching experiences. Pugh [Pug90] presents a good perspective on identifying requirements and constraints, although with more emphasis on mechanical systems. The article by Robert Abler [Abl91] is a short primer that provides good advice on how to develop specifications that overlaps with the properties presented in the IEEE Standard 1233 [IEEE Std. 1233-1998].

The HOQ technique was originally developed by Hauser and Clausing [Hau88] and has gained wide acceptance. Their original article provides a case study of the technique applied to the design of automobile door seals as implemented by Toyota Motor Corporation. Ullrich and Eppinger [Ull03] present a good perspective on developing specifications employing the QFD techniques and the HOQ with an emphasis on the voice of the customer. Ullrich and Eppinger break down the HOQ into separate matrices in a manner similar to that presented in this chapter.

3.10 Problems

3.1 Describe the relationship (similarities and differences) between *engineering requirements* and *performance requirements*.

3.2 Describe the relationship between the *system requirements specification* and *subsystem design specification*.

3.3 Describe the difference between *verification* and *validation*.

3.4 Explain how *validation* is performed for a Requirements Specification.

3.5 For each of the engineering requirements below, determine if it meets the properties of abstractness, unambiguous, and verifiable as defined in Section 3.3. If a requirement does not satisfy the properties, restate it so that it does:

    a) The TV remote control will be easy to use.

b) The robot will identify objects in its path using ultrasonic sensors.

c) The car audio amplifier will be encased in aluminum and will operate in the automobile environment.

d) The audio amplifier will have a total harmonic distortion that is less than 2%.

e) The robot will be able to move at speed of 1 foot/sec in any direction.

f) The system will employ smart power monitoring technology to achieve ultra-low power consumption.

g) The system shall be easy to use by a 12 year old.

3.6 Consider the design of a common device such as an audio CD player, an electric toothbrush, or a laptop computer (or other device that you select). Identify potential marketing and engineering requirements. Consider performance, energy, and environmental engineering requirements, as well as any others that are applicable to the problem. You do not need to select the target values, but should identify the measures and units.

3.7 Develop a marketing-engineering tradeoff matrix for the device selected in Problem 3.6.

3.8 Develop an engineering tradeoff matrix for the device selected in Problem 3.6.

3.9 Consider the devices selected in Problem 3.6. Identify the constraints that apply to the design. Address all of the eight constraint categories presented in this chapter, and in each case provide a short description as to how it does or does not apply to the problem.

3.10 Identify the three levels of standards usage and what is meant by each one.

3.11 Develop a list of potential standards that would apply to one of the devices proposed in Problem 3.6. For each case, indicate how they would apply.

3.12 **Project Application.** Develop a complete Requirements Specification document for your project that includes the requirements, tradeoffs, constraints, and standards that are applicable to the design. Make sure that the requirements, constraints, and standards are relevant for the particular problem. A format for documentation is provided in Section 3.8, along with a self-assessment checklist in Table 3.7. Also be sure to provide a justification that supports the realism of the requirements.

Chapter 4   Concept Generation and Evaluation

Creativity is a great motivator because it makes people interested in what they are doing. Creativity gives hope that there can be a worthwhile idea. Creativity gives the possibility of some sort of achievement to everyone. Creativity makes life more fun and more interesting.— Edward DeBono

When developing a design, it is important to explore many potential solutions and select the best one from them. Too often a single concept is generated and is the only one pursued, the unfortunate result being that potentially better solutions are not considered. When confronted with a problem, engineers must explore different concepts, critically evaluate them, and be able to defend the decisions that led to a particular solution. Two key thought processes employed are creativity and judgment. Creativity involves the generation of novel concepts, while judgment is applied to evaluate and select the best solution for the problem. Creativity and judgment appear to be inherent individual qualities that can't be taught. That is to some extent true, but with practice and application of formal techniques, they can be improved.

It is important to distinguish between innovation and creativity. Creativity refers to the ability to develop new ideas, while innovation is the ability to bring creative ideas to reality. Innovation is valued by companies since new products and services are often their lifeblood. That is why many make it a priority to hire engineers who can bring creativity to the design process. This chapter addresses creativity, concept generation, and evaluation in design. It presents barriers to creative thought and strategies for overcoming them. Formal approaches for generating and evaluating concepts are examined and applied to design problems.

Learning Objectives

By the end of this chapter, the reader should:

- Understand the importance of creativity, innovation, concept generation, and critical evaluation in engineering design.
- Be familiar with the barriers that hinder creativity.
- Be able to apply strategies and formal methods to generate concepts.
- Be able to apply decision-making techniques for the evaluation of design concepts.

4.1 Creativity

Is creativity something that is inherent in the individual or something that can be learned? It appears that both are true; some individuals are naturally more creative than others, yet people can enhance their creativity with conscious effort and practice. This section examines barriers to creativity, different thinking modes, and strategies for enhancing creativity. One of the ways to spark creativity is to solve puzzles. To get into the creative spirit the reader can try to solve the puzzles presented in Figure 4.1.

(a)   (b)

Figure 4.1 (a) The shovel problem. Think of this as a shovel with a coin on the spade. The objective is to move two lines so that the coin is no longer in the spade, but there is still a shovel. (b) The nine dot problem. Draw four connected straight lines that pass through all nine dots.

4.1.1 Barriers to Creativity

James L. Adams, an engineer and former professor at Stanford University, has researched innovation in technical domains. He examined the barriers to creativity and classified them into the following four types: 1) perceptual blocks, 2) emotional blocks, 3) cultural and environmental blocks, and 4) intellectual and expressive blocks [Ada01].

Perceptual blocks are those that prevent people from clearly seeing the problem for what it is. A common perceptual block is the tendency to delimit the problem space, or in other words, to put constraints on the problem that don't exist. Have you solved the puzzles shown in Figure 4.1 yet? If not, it is possible that you are placing constraints on the problems that don't exist. Knowing that this is the case, you may want to go back and try again. Another example of a perceptual block is the tendency to stereotype or see a solution to a problem that one is biased to see. This occurs because we have used similar techniques for solving the problem in the past. For example, if you have used a microcontroller to solve a certain type of problem, chances are that you are going to consider using a microcontroller in all related problems in the future. Another perceptual block is the difficulty of isolating the true problem. Three pictures that illustrate this are shown in Figure 4.2. When examining these images, people tend to form a conclusion as to what the content of each one is. Look carefully, as each picture has two equally valid interpretations.

Figure 4.2 Each of the images shown above has two different interpretations. Can you determine what they are?

One of the most common *emotional blocks* is the fear of failure. People often have creative ideas, but are afraid to express them since they may be criticized or may not have the "correct" answer. It is cliché to hear that you must fail often to succeed, but true. The highly successful product design company, IDEO, that was examined in Chapter 2 takes the approach in concept generation to "*fail early and often*" in order to succeed [Kel01]. Their design teams are encouraged to develop many seemingly outlandish ideas that are often discarded, but sometimes lead to innovative solutions. Another emotional block is a fear of chaos and disorganization. The creative process challenges engineers, since it is disorganized and not a neat scientific approach to which they are accustomed. Another block is the tendency to critically judge ideas, rather than generate and build upon them. Finally, it takes time for creative ideas to incubate. Most of us can relate to the experience when we could not solve a problem that nagged us for a period of time, followed by that unexpected "*Aha!*" moment when we identified the solution.

Environmental blocks refer to those things in our environment that limit creative ability. This could be in the form of poor teamwork where members distrust each other and criticize each other's ideas. In the workplace, this could be due to autocratic management that resists new ideas. There are also cultural biases against creativity. There is a bias against creativity as an approach to problem solving in the engineering field. This is usually based upon the reasoning that there is a single correct solution to a problem and creativity is an excuse for poor engineering. It is true that creativity and brainstorming alone do not solve engineering problems—the concepts generated need to be scrutinized using engineering principles to become viable innovations.

The final block that Adams identified is that of *intellectual and expressive*. In an engineering context, this means that the designer needs to have an understanding of intellectual tools that are applied to solve problems. For example, mathematics is a universal language for expressing and solving scientific problems. Specific examples in ECE are languages that describe the characteristics of systems such as functional, logical, and state behaviors. Examples in digital design are truth tables (input, output behavior) and state diagrams (stimulus-response). In the domain of electronics design, a functional approach (input, output, and function) is commonly used. Chapters 5 and 6 present tools for modeling the behavior of ECE systems.

4.1.2 Vertical and Lateral Thinking

Edward DeBono is the father of a field known as *lateral thinking*, which offers a different perspective on the barriers to creativity. Lateral thinking is contrasted to what is known as the vertical thinking process [Deb67, Deb70]. Engineers tend to be vertical (or convergent) thinkers, meaning that they are good at taking a problem and proceeding logically to the solution. This is typically a sequential linear process, where the engineer starts at the highest level and successively refines elements of the design to solve the problem. This is usually based upon experience solving similar problems and conventional tools that are employed in that particular area.

The objective of lateral (or divergent) thinking is to identify creative solutions. It is not concerned with developing the solution for the problem, or right or wrong solutions. It encourages jumping around between ideas. In the words of DeBono *"The vertical thinker says: 'I know what I am looking for.' The lateral thinker says: 'I am looking but I won't know what I am looking for until I have found it.'"* The field of lateral thinking is characterized by puzzles of the following type found at Paul Sloane's Lateral Thinking Puzzles website [http://dspace.dial.pipex.com/sloane]:

> *A body is discovered in a park in Chicago in the middle of summer. It has a fractured skull and many other broken bones, but the cause of death was hypothermia.*

> *A hunter aimed his gun carefully and fired. Seconds later, he realized his mistake. Minutes later, he was dead.*

> *A man is returning from Switzerland by train. If he had been in a non-smoking car he would have died.*

The objective in these puzzles is to develop plausible scenarios that explain how each of the above situations could have happened. A solution for the first example is that a person stowed away in the wheel compartment of a jet airliner. While in flight he froze and died of hypothermia. When the plane prepared to land, it lowered its landing gear, causing the body to fall to the park, fracturing his skull, and breaking his bones. Can you develop plausible scenarios that describe each of them?

Vertical thinking focuses on sequential steps toward a solution and tries to determine the correctness of the solution throughout the process. This is very different from lateral thinking where there is nonlinear jumping around between steps and there is no attempt to discern between right and wrong. As such, lateral thinking is more apt to follow least likely paths to a solution, whereas vertical thinking follows the most likely paths. The goal in lateral thinking is to develop as many solutions as possible, while vertical thinking tries to narrow to a single solution.

Lateral thinking is appropriate for the concept generation phase. So should concept generation and brainstorming be done by the individual or by a team? DeBono and Osborn [Osb63] conclude that creativity is more effective by individuals than by teams. However, Osborn also points out that there is great value in applying creativity in teams, since it provides a place for the team to work together on problems and see other perspectives. Our anecdotal observations of student design teams supports this—group brainstorming is effective for developing concepts, new product ideas, new features, and different ways to combine technologies. This is because in groups, ideas are readily built upon by other team members. More mathematical, technical, and theoretical breakthroughs tend to be the work of the lone genius. Examples of this are the Theory of Relativity (Einstein), Boolean Logic (Bool), and Shannon's Sampling Theorem. We have also observed that novice designers, who do not have much experience in concept generation, can benefit greatly from group brainstorming techniques.

4.1.3 Strategies to Enhance Creativity

There are valuable strategies that can be employed to enhance the creative process. The body of research on the subject is very large and some key points are summarized as follows:

- *Have a questioning attitude.* One of the keys is to have a questioning attitude and challenge assumptions. The willingness to do this generally decreases as people age. Young children are highly creative and are constantly questioning everything, with questions such as *"Why do trees have leaves?"* or *"Why is the sky blue?"* Asking basic questions stimulates creativity and is applicable to technical designs. When examining a design with a microcontroller, ask questions such as *"Is there a way to replace the microcontroller?"*, *"Are there other features that I can achieve with the microcontroller?"*, and *"Is there a better microcontroller that can be used?"*

- *Practice being creative.* Research shows that people can improve their creative ability through conscious effort. For example, try solving the puzzles presented in this chapter and in the end of the chapter problems. Be conscious of things that bother you ("pet peeves") in your everyday life and try to develop new solutions for them.

- *Suspend judgment.* It is easy to criticize and immediately dismiss ideas, so it is important to defer judgment and be flexible in thinking. Seemingly outlandish ideas can lead to other concepts that are valuable solutions. The opportunity for new solutions is curtailed if ideas are immediately judged and discarded. Creative concepts can be developed by taking a concept and modifying it or combining it with other seemingly unrelated concepts.

- *Allow time.* The creative process needs time for incubation. The human mind needs time to work on problems, so set aside time to reflect on the problem and to allow it to incubate so that the *"Aha!"* moment of discovery can happen.

- *Think like a beginner.* New solutions often come from novices. The reason is that novices don't have preconceived ideas as to the solution for a problem. Experience is a double-edged sword—it allows one to quickly solve problems by drawing upon pre-existing solutions, but can inhibit creativity. If confronted with a new problem that bears similarity to one encountered in the past, then it is likely that the new solution will bear similarity to the old one. If everyone else is it doing it one way, consider the opposite.

4.2 Concept Generation

Many creative ideas arise from novel combinations and adaptations of existing technology. SCAMPER, an acronym for Substitute, Combine, Adapt, Modify, Put to other use, Eliminate, and Rearrange/Reverse, can be used as a guide to systematically generate creative concepts. Each element is described via the questions presented below:

- *Substitute.* Can new elements be substituted for those that already exist in the system?
- *Combine.* Can existing entities be combined in a novel way that has not been done before?
- *Adapt.* Can parts of the whole be adapted to operate differently?
- *Modify.* Can part or all of a system be modified? For example, size, shape, or functionality.
- *Put to other use.* Are there other application domains where the product or system can be put to use?
- *Eliminate.* Can parts of the whole be eliminated? Or should the whole itself be eliminated?
- *Rearrange or Reverse.* Can elements of the system be rearranged differently to work better? This is different from substituting in that the elements of the system are not changed, but rearranged or ordered differently to create something new. In terms of reversal, are there any roles or objectives that can be reversed?

SCAMPER is a modification of a set of questions that was originally posed by Osborn [Osb63] and was modified to its form above by Michalko [Mic91].

A *concept table*, demonstrated in Table 4.1, is a tool for systematically examining different combinations, arrangements, and substitutions. The headings in the table identify characteristics to be achieved in the design, while the entries in the corresponding column represent potential solutions. Novel products or solutions are generated by combining elements from each of the columns, which are identified in the table by circled elements. The solutions can be in the form of a single element selected from each column, or as in the example shown, multiple elements selected from each column.

Table 4.1 A concept table for generating ideas for a personal computing system. The potential solution is identified by the combination of circled elements.

User Interface	Display	Connectivity & Expansion	Power	Size
Keyboard	CRT	Serial & Parallel	Battery	Hand-held, Fits in pocket
Touchpad	Flat Panel	USB	AC Power	Notebook size
Handwriting Recognition	Plasma	Wireless Ethernet	Solar Power	Wearable
Video	Heads-up Display	Wired Ethernet	Fuel Cell	Credit Card Size
Voice	LCD	PCMCIA	Thermal Transfer	Flexible in shape
		Modem / Telephone		

Based on the concepts circled in Table 4.1, one can imagine a personal computing system that has the following features: 1) is wearable with different credit card size components placed on the body and in clothing to make it comfortable to use; 2) is powered by a combination of solar cells, fuel-cells, and from thermal heat generated by a person's body; 3) has a microphone and camera integrated in the user's clothing for interface to the system, as well as a flexible foldable keyboard for typing that is stored in a pocket; 4) has a heads-up display integrated with the user's eyeglasses or baseball hat; and 5) has a miniature earpiece microphone used for communication. While the above example focused on novel combinations and substitutions, the concept table can also be used to examine the possibility of eliminating ideas. For example, the table inherently assumes that a display will be used. However, it should also be asked if it is absolutely necessary in the design.

Another example is shown in Table 4.2, where the objective is to identify design concepts for a temperature measurement and display device. There are three main elements to the proposed solution: the thermal sensing method, circuitry that converts the sensor information (temperature) to a voltage, and a display unit that converts the voltage to a displayed temperature. Note that the table implies a three-stage architecture, thus concepts are generated within that framework. There may be completely different architectures that are better.

Table 4.2 Concept table for a temperature measurement device.

Thermal Sensing	Conversion to Voltage	Display
Thermistor	Op Amp Design	Seven-Segment LEDs
RTD	Transistor Designs	LCD
Thermocouple		Analog Dial Indicator

A related tool is a *concept fan,* which is a graphical representation of design decisions and choices. An example concept fan for the temperature measuring device is shown in Figure 4.3. Design decisions are identified by circles; solutions are indicated by squares. In this example, more options are shown than in Table 4.2. Concepts are generated by selecting among the different solution blocks.

Concept generation in teams is also fostered using formal methods, with the most well-known being *brainstorming*. Group brainstorming is effective for generating many concepts in a short period of time. Experienced design teams are known to generate hundreds of concepts in an hour. Traditional brainstorming is not highly-structured—though a facilitator helps—and employs five basic rules:

- No criticism or judgment of ideas.
- Wild ideas are encouraged.
- Quantity is stressed over quality.
- Build upon and modify the ideas of others.
- All ideas are recorded.

Another group technique is known as the *Nominal Group Technique* (NGT) [Del71]. In NGT each team member silently generates ideas that are reported out in a round-robin fashion so that all members have an opportunity to present their ideas. Concepts are selected by a multi-voting scheme with each member casting a predetermined number of votes for the ideas

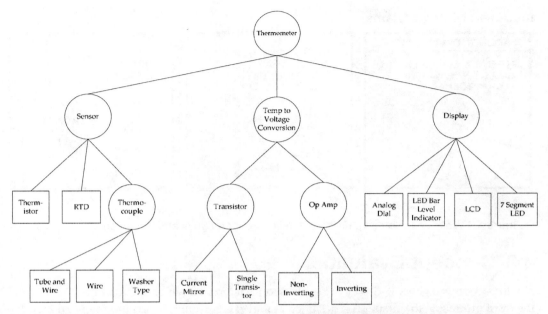

Figure 4.3 A concept fan for the temperature measuring device. The circles represent the choices to be made and the squares represent potential solutions to the choices.

presented. The ideas are then ranked, discussed further, and voted upon again if necessary. The steps of NGT are as follows:

1. *Read problem statement.* It should be read out loud by a team member (the facilitator).

2. *Restate the problem.* Each person restates the problem in their own words to ensure that all members understand it.

3. *Silently generate ideas.* All members silently generate ideas during a set period of time, typically 5–15 minutes.

4. *Collect ideas in a round-robin fashion.* Each person presents one idea in turn until all ideas are exhausted. The facilitator should clarify ideas and all should be written where the entire team can view them.

5. *Summarize and rephrase ideas.* Once the ideas are collected, the facilitator leads a discussion to clarify and rephrase the ideas. This ensures that the entire group is familiar with them. Related ideas can be grouped or merged together.

6. *Vote.* Each person casts a predetermined number of votes, typically three to six, for the ideas presented. The outcome is a set of prioritized ideas that the team can further discuss and pursue.

DILBERT® by Scott Adams

Figure 4.4 Wally brainstorming. (Dilbert © United Feature Syndicate. Reprinted by permission.)

4.3 Concept Evaluation

After concepts are generated they are evaluated to determine which are the most feasible and the most promising to pursue. The designer should exercise engineering judgment and use the customer needs and technical factors to drive the decision. This process is shown in Figure 4.5, where the user needs, potential solutions, and technical factors serve as inputs to a decision process that ranks the concepts. A point of caution—some of the methods presented generate numerical scores for comparing concepts, potentially leading one to believe that the quantitative results are infallible. Keep in mind that the inputs are based on qualitative and semi-quantitative assessments and can be geared to select a preconceived notion of the solution. It is important to maintain flexibility of thinking, to challenge assumptions, and ultimately select the best concept. Several methods of evaluation are presented.

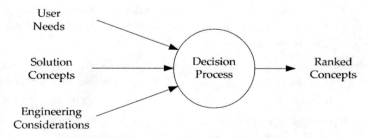

Figure 4.5 Process for concept evaluation.

4.3.1 Initial Evaluation

The concepts generated should be initially reviewed and those that are completely infeasible discarded. Some of the reasons a concept may be deemed infeasible are that it may be far too costly, will take too long to develop, or involve too much risk. In many cases it may be deemed that using cutting-edge technology represents an unacceptable risk. Concepts that clearly cannot meet the engineering requirements should also be discarded. Care should be taken not to completely eliminate ideas that may have merit, as conditions change and some concepts that were previously thought unrealistic may become viable in the future.

4.3.2 Strengths and Weaknesses Analysis

Another form of evaluation is to complete a *strengths and weaknesses analysis* of the potential solutions. Table 4.3 demonstrates the application of this analysis applied to an experimental design project for testing of the Intel 1000XF card (examined in Chapter 2 (Example 2.2) and Chapter 3 (Table 3.2)). In order to test the card under different operating temperatures, a method of heating the card and holding its temperature fixed during the experiment was needed. The two solutions compared were to use a contact heating element or to place the card in an environmental test chamber. In this particular example, the temperature chamber solution was ultimately selected due to the need for a uniform temperature distribution. The strength and weakness analysis is good for examining problems of moderate complexity. It suffers in that it does not require uniform criteria for comparison and is not quantitative.

Table 4.3 A strengths and weaknesses analysis of proposed methods for heating an Intel 1000XF card to be used in lifetime testing. [Ese03].

Method	Strengths	Weaknesses
Contact Heating	• Simplest design • Could be used internally to computer	• Does not create uniform temperature • Hard to control temperature
Temperature Chamber	• Uniform temperature • Greater control over temperature	• Must be external to computer • More difficult to design • Expensive

4.3.3 Decision Matrices

Decision matrices integrate technical knowledge and the needs of the end-user to generate a numerical score for comparing concepts. The idea is to compare design alternatives against pre-selected criteria. Typically, the criteria are the engineering or marketing requirements, and they should be selected before any evaluation begins. If not, it is possible that one concept, most likely the favorite, may be used to develop the comparison criteria.

In its simplest form, a decision matrix is similar to the benchmarking presented in Chapter 3 where each design option is compared against the criteria. The scale used for comparison is flexible and one example is to use a [1–5] rating scale where [1] = does not meet criterion, [3] = partially meets criterion, and [5] = completely satisfies the criterion. A hypothetical example is shown in Table 4.4. Summing the scores in each column indicates that the second design option is the best. The advantage of this relative to the strengths and weaknesses analysis is that the same criteria are used for all options and a consistent rating is used. It does not take into account that the comparison criteria may have different weightings.

Table 4.4 A simple decision matrix.

	Design Concept 1	Design Concept 2	Design Concept 3
Criteria 1	2	5	4
Criteria 2	1	3	2
Criteria 3	3	1	1
Criteria 4	4	4	3
Score	10	13	10

A more complex and quantitative decision matrix is shown in Table 4.5. The criteria are listed in the leftmost column, while the associated weighting factors (ω_i) quantify the relative importance of the criteria. The body of the matrix contains design ratings, α_{ij}, that reflect the technical merit of each of the jth design options relative to i^{th} criterion. The total score, S_j, for each design option is computed as a weighted summation of the design ratings and weighting factors as follows

$$S_j = \sum_{i=1}^{m} \omega_i \alpha_{ij} \,. \tag{1}$$

Table 4.5 A quantitative decision matrix.

		Design Option 1	Design Option 2		Design Option n
Criteria 1	ω_1	α_{11}	α_{12}	$\cdots$	α_{1n}
Criteria 2	ω_2	α_{21}	α_{22}	$\cdots$	α_{2n}
$\vdots$	$\vdots$	$\vdots$	$\vdots$	$\cdots$	$\vdots$
Criteria m	ω_m	α_{m1}	α_{m2}	$\cdots$	α_{mn}
Score		$S_1 = \sum_{i=1}^{m} \omega_i \alpha_{i1}$	$S_2 = \sum_{i=1}^{m} \omega_i \alpha_{i2}$	$\cdots$	$S_n = \sum_{i=1}^{m} \omega_i \alpha_{in}$

The steps applied to develop a decision matrix are now examined in more detail.

Step 1: Determine the Criteria

The criteria are typically determined from the engineering and/or marketing requirements and are selected *a priori* to comparison of the concepts.

Step 2: Select the Weighting Factors (ω_j)

The weighting factors represent the relative importance of the comparison criteria. The choice of scale is flexible and is selected based upon the particular application, but it is important to be consistent and use the same scale for all weighting factors. One possibility is to assign numerical values based on a subjective ranking, such as

[1] = optional, [3] = important, [5] = critical.

Another possibility is to employ the pairwise comparison results determined in Chapter 2. In pairwise comparison each criterion is compared against all others and the output is a score, ρ, for each criterion. The weighting factor is determined as a ratio of the individual score divided by the sum of scores for all criteria as

$$\omega_j = \frac{\rho_j}{\sum_k \rho_k} . \tag{2}$$

Dividing by the sum of scores normalizes the weightings and ensures consistency in the comparison.

Step 3: Select the Design Ratings (α_{ij})

The design ratings are scored on the technical merits of the design and are the main entries in body of the table. An example scoring system is a [1–5] scale that indicates how well the design meets the criteria, such as, [1] = does not meet criterion at all, [3] = partially meets criterion, [5] = completely satisfies the criterion.

Quantitative ratings are better if they can be obtained, and sometimes require a little creativity to produce. For example, assume that one of the comparison criteria is cost, and the objective is to minimize it. A simple design rating for cost is

$$\alpha = \frac{\min\{cost\}}{cost} . \tag{3}$$

This scoring technique assigns the highest score of 1 to the minimum cost value. In cases where the objective is to maximize a particular quantity, a similar measure can be used where the ratio is inverted and the min{} operator is replaced by the max{} operator. The decision ratings in each row (for a particular criterion) are normalized so that the sum of the α values in each row is equal to one. This ensures that each criterion is treated equally with the weighting controlled only by the ω_i values.

Step 4: Compute the Scores (S_j)

At this point the entries for the matrix are known and the scores for the alternatives are computed from the weighted sum in (1).

Step 5: Review the Results

The output is a set of numerical scores that rank the design options. Close contenders should be reviewed to determine the differences between them and the highest scoring option, and whether the differences are significant. Remember that the inputs to the process are subjective. The final result should be reviewed to see if it makes sense. Keep the following quote in mind when interpreting the results.

> *You can use all the quantitative data you can get, but you still have to distrust it and use your own intelligence and judgment.*—Alvin Toffler

Example 4.1 demonstrates the application of the decision matrix technique.

Example 4.1 Selection of a current source for temperature measurement.

Problem: Consider the design of an electronic circuit for measuring temperature where the objective is to create a constant current source that drives a temperature measurement device (RTD). The resistance of an RTD varies with temperature, and when driven by a constant current, I, produces a voltage, V_T, that varies proportionally with temperature. Assume that the objective is to supply a constant current of $I=1mA$ and that the three circuits below are being considered.

1: Single Transistor   2: Inverting Op Amp   3: Current Mirror

Assume the following information is also known:

- The comparison criteria for the design are: 1) high accuracy, 2) low cost, 3) small size, and 4) availability for manufacture. Assume that they were ranked using a pairwise comparison (from Chapter 2) as shown below.

	Accuracy	Cost	Size	Availability	Sum
Accuracy	-	1	1	0	2
Cost	0	-	0	½	½
Size	0	1	-	0	1
Availability	1	½	1	-	2½

- The parts costs are the following: resistors = $0.05, bipolar junction transistors (BJTs) = $0.15, op amps = $0.35, and RTDs = $0.25.

- The accuracy of each design was evaluated by a sensitivity analysis using a circuit simulation package assuming 10% resistors. The deviation of the output voltage (maximum deviation from nominal) for the three designs is 9.2%, 1.3%, and 1.9% respectively.

- To manufacture each circuit on a printed circuit board requires the following dimensions: design 1 = 1 in², design 2 = 1.56 in², and design 3 = 2.25 in².

- The parts have an in-stock availability of 95%, 70%, 90%, and 80% of the time for the resistors, BJTs, RTDs, and op amps respectively.

The problem is to apply a weighted decision matrix to rank the design concepts.

Solution:

Step 1: Determine the Criteria
In this example the comparison criteria of accuracy, cost, size, and availability were given as part of the problem.

Step 2: Select the Weighting Factors
The weighting factors for the criteria are selected based upon the scores from the pairwise comparison. Normalizing the weighting factors as indicated in (2) produces $\omega_1 = 2/6 = 0.33$, $\omega_2 = 0.5/6 = 0.08$, $\omega_3 = 1/6 = 0.17$, and $\omega_4 = 2.5/6 = 0.42$.

Step 3: Select the Design Ratings
Design ratings need to be made for each of the criterion.

Accuracy. Since the objective is to minimize the deviation, the following rating is used

$$\alpha = \frac{\min\{deviation\}}{deviation} \ .$$

This produces the following design ratings for accuracy: $\alpha_{11} = 0.14$, $\alpha_{12} = 1.0$, and $\alpha_{13} = 0.68$; and normalizing the row sum produces $\alpha_{11} = 0.08$, $\alpha_{12} = 0.55$, and $\alpha_{13} = 0.37$.

Cost. Using the cost measure in (3) gives following cost ratings for the three options respectively: $\alpha_{21} = 1.0$, $\alpha_{22} = 0.69$, and $\alpha_{23} = 0.75$. Normalizing the row sum produces $\alpha_{21} = 0.41$, $\alpha_{22} = 0.28$, and $\alpha_{23} = 0.31$.

Size. The objective is to minimize size, and using a measure analogous to (3) for the required space to manufacture each produces the following normalized decision ratings: $\alpha_{31} = 0.48$, $\alpha_{32} = 0.31$, and $\alpha_{33} = 0.21$.

Availability. A measure for the overall availability of parts to manufacture each design is required. One way to measure this is to compute the probability that a design will be able to be manufactured based upon the past history of part availability. This is found by multiplying the availability of all individual components needed for the design:

$$P(\text{design 1 can be produced}) = (0.95)(0.90)(0.70) = 0.60$$
$$P(\text{design 2 can be produced}) = (0.95)(0.90)(0.80) = 0.68$$
$$P(\text{design 3 can be produced}) = (0.95)(0.90)\ (0.70)(0.70) = 0.42$$

This produces the following normalized decision ratings for availability: $\alpha_{41} = 0.35$, $\alpha_{42} = 0.40$, and $\alpha_{43} = 0.25$.

Step 4: Compute the Scores

		Single BJT	Op Amp	Current Mirror
Accuracy	0.33	0.08	0.55	0.37
Cost	0.08	0.41	0.28	0.31
Size	0.17	0.48	0.31	0.21
Availability	0.42	0.35	0.40	0.25
Score		0.29	0.42	0.29

Step 5: Review the Decision

Remember that this is a semi-quantitative method. The final ranking indicates that design options 1 and 3 are essentially equal, while both are inferior to option 2.

4.4 Project Application: Concept Generation and Evaluation

Design teams should set aside time for brainstorming and concept generation. Brainstorming can be free-form generation of ideas following the rules of brainstorming provided or carried out more formally using the Nominal Group Technique. Designers should take it as a challenge to identify many different concepts for solving problems. Remember that effective design teams can generate hundreds of concepts in a brainstorming session. Concepts can be generated for the entire design as well as for individual components of the design. The techniques in this chapter can be combined with the design methods presented in Chapters 5 and 6 to evaluate design concepts. SCAMPER, concept tables, concepts fans, and decision matrices can be used to document the results of the team's concept generation and evaluation. Ultimately the team must use judgment to select and apply tools that are relevant to the problem.

4.5 Summary and Further Reading

In the design process, it is important to creatively generate different concepts for a solution to the problem. This is followed by an evaluation of concepts to determine which are the most promising. This chapter identified barriers to creativity and provided strategies for enhancing creative ability. The concepts of vertical and lateral thinking were introduced, and their impact on the design process was explored. Formal methods of concept generation, including concept tables, concepts fans, and group brainstorming techniques, were presented. Finally, methods for critically evaluating concepts were presented and applied to a design problem.

There are many references that examine creativity and concept generation. Adams [Ada01] is a good reference for creativity and problem-solving with a technical bent. Alex Osborn was an advocate of creativity and developed two readable works that address the creative process, the need for creativity, and strategies for enhancing it [Osb48, Osb63]. Edward DeBono is another well-known authority in the field and has produced many works on lateral thinking and creativity [Deb67, Deb70]. Paul Sloane has published numerous books with lateral thinking puzzles [Slo91, Slo93, Slo94].

Several techniques for concept generation were presented and there are other reported approaches such as TRIZ [Alt99], which is a systematic approach to technical innovation. There has also been much work done in the field of decision making, and one that follows the techniques presented here is the Analytical Hierarchy Process (AHP) [Sat88]. It is similar to the decision matrix technique with weighting factors and technical ratings, but also attempts to reduce bias in the ratings. Another common approach applied for design evaluation is Pugh Concept selection, in which concepts are iteratively compared [Pug90].

4.6 Problems

4.1   Consider the nine dot puzzle shown in Figure 4.1 (b). Draw **three** connected straight lines that pass through all nine dots.

4.2   Consider the six sticks shown below. Rearrange the sticks to produce four equilateral triangles (the sticks cannot be broken).

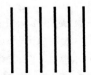

4.3   Consider the fish shown below made of eight sticks and a coin for the eye. The objective is to make the fish face the other direction by moving only the coin and three sticks.

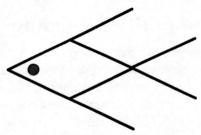

4.4   For each of the following lateral thinking puzzles, develop a plausible solution (from Paul Sloane's <u>Lateral Thinking Puzzles</u> [http://dspace.dial.pipex.com/sloane]):

   a)   A man walks into a bar and asks the barman for a glass of water. The barman pulls out a gun and points it at the man. The man says 'Thank you' and walks out.

   b)   A woman had two sons who were born on the same hour of the same day of the same year. But they were not twins. How could this be so?

   c)   Why is it better to have round manhole covers than square ones?

   d)   A man went to a party and drank some of the punch. He then left early. Everyone else at the party who drank the punch subsequently died of poisoning. Why did the man not die?

4.5   Legislation was passed to allow handguns in the cockpits of passenger airliners to prevent hijacking. Brainstorm to develop concepts that prevent anyone other than the pilot from using the handgun.

4.6   Imagine if scientists and engineers were able to develop a technology that would allow people to be transported from any place on earth to another instantaneously. Brainstorm to determine the potential impact this would have on society.

4.7   Student advising at many colleges and universities is seen as an area that can be improved. Brainstorm to develop ideas as to how student advising could be improved at your college or university.

4.8   In your own words, describe what a concept table and a concept fan are.

4.9   Consider Example 4.1 in the chapter. For this example assume that:

- The following is the result of the paired comparison.

	Accuracy	Cost	Size	Availability	Sum
Accuracy	-	0	1	0	
Cost	1	-	1	½	
Size	0	0	-	0	
Availability	1	½	1	-	

- The parts costs are the following: resistors = $0.05, bipolar transistors (BJTs) = $0.10, op amps = $0.35, and RTDs = $0.25.

- The parts have an in-stock availability of 99%, 90%, 85%, and 70% of the time for the resistors, BJTs, RTDs, and op amps respectively.

- Everything else is the same as presented in Example 4.1.

Compute the rankings of the design options using a weighted decision matrix of the type shown in Table 4.5.

4.10   **Project Application.** Conduct brainstorming to identify concepts for solving the design problem. Critically evaluate the concepts generated using one or more of the techniques presented in the chapter that is appropriate for the problem. Section 4.4 provides guidance on how to conduct this and document the results.

Part II – Design Tools

Chapter 5   System Design I: Functional Decomposition

At Sony, we assume all products of our competitors will have basically the same technology, price, performance, and features. Design is the one thing that differentiates one product from another in the marketplace. —Norio Ohgo, Chairman and CEO, Sony

After the technical concept is selected, it is translated into a solution that satisfies the system requirements. The designer must put on paper, or the computer screen, a representation that is meaningful and clear; in other words, a useful abstraction of the system. Engineering designs are often complex, consisting of many systems and subsystems, thus this representation should facilitate the design process and effectively describe the system. In addition, it serves an important function in communicating the design to all members of the team. Imagine a scenario where each team member is responsible for designing part of a large system. Each person develops their part in isolation and several months later the team gets back together to integrate the pieces. Of course, the system won't work unless the team has collectively defined and communicated the functionality and interfaces for all subsystems in the design.

This chapter presents a well-known design technique—known as *functional decomposition*—that is intuitive, flexible, and straightforward to apply. It is probably the most pervasive design technique used for engineering systems and is applicable to a wide variety of problems that extend well beyond ECE. In functional decomposition, systems are designed by determining the overall functionality and then iteratively decomposing it into component subsystems, each with its own functionality.

The objective of this chapter is to present both basic design concepts and the functional decomposition design technique. A process for functional decomposition is provided and it is applied to examples in analog electronics, digital electronics, and software systems.

Learning Objectives

By the end of this chapter, the reader should:

- Understand the differences between bottom-up and top-down design.
- Know what functional decomposition is and how to apply it.
- Be able to apply functional decomposition to different problem domains.
- Understand the concepts of coupling and cohesion, and how they impact designs.

5.1 Bottom-Up and Top-Down Design

Two general approaches to synthesizing engineering designs are known as *bottom-up* and *top-down*. In the case of bottom-up, the designer starts with basic components and synthesizes them to create the overall system. To use an analogy, consider the case of creating an automobile. In the bottom-up approach, you have pieces of the automobile, such as the tires, motor, axle, transmission, alternator, and they are brought together to create a car. The implication is that the final system depends upon the parts at hand. In other words, in the bottom-up approach, the parts and subsystems are given, and from them an artifact is created.

The top-down approach is analogous to the concept of divide-and-conquer. In top-down the designer has an overall vision of what the final system must do, and the problem is partitioned into components, or subsystems that work together to achieve the overall goal. Then each subsystem is successively refined and partitioned as necessary. In the case of the automobile, the overall objective is determined; the major subsystems are defined, such as electrical, power drive-train, and the suspension; and then each subsystem is further refined into its component parts until the complete system is designed.

A debate that continues in the design community revolves around which is the better approach. The obvious answer is that top-down appears to be better, since it starts with the overall goal (requirements) and from that a solution is developed. Top-down is particularly valuable on large projects with many subsystems, where it is unlikely that bringing together pieces in an ad-hoc fashion will successfully solve the problem. A disadvantage of top-down design is that it tends to limit the solution space and innovation. Top-down design is inclined to follow a vertical thought process (Chapter 4) where the designer starts with a problem and successively refines the subsystems until a blueprint for solving the problem is defined. Furthermore, the designer cannot create a top-down design in a vacuum without bottom-up knowledge of existing technology and how the system can be realized.

Bottom-up has the advantage of lending itself to creativity. It allows the designer to take different technologies and from them create something new, allowing more *"what if?"* questions to be asked. Bottom-up design is applicable when there are constraints on the components that can be used. This is a realistic scenario. Consider the case of variant design, where the goal is to improve the performance of an existing, or legacy, system. For example, automobile manufacturers might have to redesign their models to meet new emissions, mileage, or safety standards. If you are not starting with a new design and must utilize existing systems, it requires bottom-up thinking. In reality, most problems require a combination of bottom-up and top-down thinking, and the designer must alternate between them.

In summary, it is most effective to work between bottom-up and top-down. A completely top-down approach is not feasible because the designer must have an understanding of the bottom level technology for the components of the design hierarchy to be realistic. Likewise, completely bottom-up by itself is generally not feasible, particularly as the system complexity grows.

5.2 Functional Decomposition

Functional decomposition is a recursive process that iteratively describes the functionality of all system components. It is analogous to the mathematical concept of a function, for example, $y=f(x)$. In this function there is an input, x, an output, y, and a transformation between the input and output, $f()$. This is easily extended to the case of multiple inputs and outputs where the inputs and outputs are vectors, $\bar{y} = f(\bar{x})$. In functional decomposition, the same elements are defined as in the mathematical analogy—the inputs, the outputs, and the transformation between the inputs and outputs (the functionality). Those three elements constitute what is known as the *functional specification* or *functional requirement*—the requirement that a functional module should meet. A *module* is a block, or subsystem, that performs a function. Functional decomposition has a strong top-down flavor, due to the fact that the highest level functionality is defined and then further refined into sub-functions, each with its own inputs, outputs, and functionality. The process is repeated until some base level functionality is reached where the modules can be actualized with physical components.

A process for applying functional decomposition is illustrated in Figure 5.1. It starts with a definition of the highest level (Level 0) of system functionality (the functional requirement for the system). This is followed by definition of the next level of the hierarchy that is needed to achieve the design objective. The Level 1 design is typically referred to as the main *design architecture* of the system. In this context, architecture means the organization and interconnections between modules. Care must be taken at each design level to ensure that it satisfies the requirements of the higher level. The process is repeated for successive levels of the design and stops when the *detailed design* level is reached. Detailed design is where the problem can be decomposed no further and the identification of elements such as circuit components, logic gates, or software code takes place. The number of levels in the design hierarchy depends upon the complexity of the problem.

Figure 5.1 A process for developing designs using functional decomposition.

5.3 Guidance

The following guidance is provided before examining applications of the functional decomposition technique:

- *It is an iterative process.* During the first pass, it is not possible to know all of the detailed interfaces between components and the exact functionality of each block. In fact, some details are not known until the implementation level is reached, so the designer needs to iterate, work between top-down and bottom-up, and adjust the design as necessary.

- *Set aside sufficient time to develop the design.* This is a corollary to the previous point. The iterative nature means that it takes time to examine different solutions and to refine the details into a working solution.

- *Pair together items of similar complexity.* Modules at each level should have similar complexity and granularity.

- *A good design will have the interfaces and functionality of modules well-defined.* It is fairly easy to piece together some blocks into an apparently reasonable design. However, the functional requirements should be clearly defined and the technical feasibility understood. If not, the design will fall apart when it comes to the implementation stage. Consider the following advice of a well-known architectural designer:

  The details are not the details. They make the design. — Charles Eames

- *Look for innovations.* Top-down designs tend to follow a vertical thinking process, where the designer proceeds linearly from problem to solution. Try to incorporate lateral thinking strategies from Chapter 4 and examine alternative architectures and technologies for the solution.

- *Don't take functional requirements to the absurd level.* Common elements, such as analog multipliers or digital logic gates, do not require explicit functional specifications. Doing so may become cumbersome and add little to the design. However, it depends upon the level at which you are working. If the goal is to design an analog multiplier chip, it is entirely appropriate to develop the functional requirement for the multiplier.

- *Combine functional decomposition with other methods of describing system behavior.* There is no single method or unifying theory for developing designs. Functional decomposition alone cannot describe all system behaviors. It may be supplemented by other tools such as flowcharts (logical behavior), state diagrams (stimulus-response), or data flow diagrams. In the digital stopwatch example presented later in this chapter, the behavior is defined using state diagrams. Other methods for describing system behavior are addressed in Chapter 6.

- *Find similar design architectures.* Determine if there exist similar designs and how they operate. Realize that this creates a bias towards existing solutions.

- *Use existing technology.* Many designers take the attitude that they are going to develop the entire design themselves, the sentiment being to ignore technology that they did not develop. Furthermore, engineering education predisposes us to design at a fundamental level. Both factors lead to time spent re-inventing the wheel. If existing technology is available that meets both the engineering and cost requirements, then use it.

- *Keep it simple.* Do not add complexity that is not needed.

  A designer knows that he has achieved perfection not when there is nothing left to add, but when there is nothing left to take away. —Antoine de St-Exupery

- *Communicate the results.* It is important to describe the theory of operation (the *why*) as well as the implementation (the *what*). The *what* in the completed design is usually quite clear from the implementation, but documenting the description of operation and design decisions helps later when the system must upgraded. Designs can also become very complex, so consider how much information can be effectively communicated on a single page. If the information is too complex to show reasonably on a page or two, then it probably is too detailed and another level in the hierarchy should be added.

5.4 Application: Electronics Design

We now examine the application of functional decomposition in different problem domains. In the domain of analog electronics, the inputs and outputs of modules are voltage and current signals. Typical transformations applied to the inputs are alterations in amplitude, power, phase, frequency, and spectral characteristics. Consider the design of an audio power amplifier that has the following engineering requirements.

The system must

- Accept an audio input signal source with a maximum input voltage of 0.5V peak.
- Have adjustable volume control between zero and the maximum volume level.
- Deliver a maximum of 50W to an 8Ω speaker.
- Be powered by a standard 120V 60Hz AC outlet.

Level 0

The Level 0 functionality for the amplifier is shown in Figure 5.2, which is fairly simple—the inputs are an audio signal, volume control, and wall outlet power, and the output is an amplified audio signal.

Figure 5.2 Level 0 audio power amplifier functionality.

The system should be described in as much detail as possible for each level via the functional requirement. The Level 0 functional requirement for this design is as follows.

Module	Audio Power Amplifier
Inputs	- Audio input signal: 0.5V peak. - Power: 120 volts AC rms, 60Hz. - User volume control: variable control.
Outputs	- Audio output signal: ?V peak value.
Functionality	Amplify the input signal to produce a 50W maximum output signal. The amplification should have variable user control. The output volume should be variable between no volume and a maximum volume level.

Not all values can be known on the first pass through the design as was indicated in the guidelines. Underlined items represent values that need to be determined or refined as the design proceeds. In this case, the peak value of the audio output voltage is determined from the system requirements on power gain. Knowing that the maximum power is given by $P_{max} = V_{peak}^2 / R$, allows the maximum output voltage to be computed as $V_{peak} = \sqrt{8\Omega * 50W} = 20V$.

Level 1

The Level 1 diagram, or system architecture, is shown in Figure 5.3. This architecture is common in amplifier design and is but one possible solution. It contains three cascaded amplifier stages and a DC supply that powers the three stages. The first amplifier stage, the *buffer amplifier*, provides a high resistance buffer that minimizes loading effects with the source. Buffer amplifiers have extremely high input resistance and a unity signal gain. The *high gain amplifier* increases the amplitude of the signal, but provides little in terms of the output current necessary to drive the speakers. The last stage in the cascade is the *power output stage*, which provides the current needed to drive the speakers, but has no voltage amplification.

Figure 5.3 Level 1 audio amplifier design.

The functional requirements for the Level 1 subsystems are now detailed, starting with the buffer amplifier.

Module	Buffer Amplifier
Inputs	- Audio input signal: 0.5V peak. - Power: ± 25V DC.
Outputs	- Audio signal: 0.5V peak.
Functionality	Buffer the input signal and provide unity voltage gain. It should have an input resistance >1MΩ and an output resistance <100Ω.

Where did the ± 25V DC value for the DC input power come from? The system must produce a ± 20V AC output signal to satisfy the Level 0 requirement, so supply values that exceed that are required to power the electronics. How about the values for the input and output resistance? They are educated guesses, based on knowledge of what is achievable with the technology (bottom-up knowledge). The exact resistance requirements are refined later based upon the overall design, taking into account the input and output resistances for all stages.

Now consider the functional requirements for the high gain amplifier.

Module	High Gain Amplifier
Inputs	- Audio input signal: 0.5V peak. - User volume control: variable control. - Power: ± 25V DC
Outputs	- Audio signal: 20V peak.
Functionality	Provide an adjustable voltage gain, between 1 and 40. It should have an input resistance >100kΩ and an output resistance <100Ω.

The gain of 40 is determined from the overall system power and gain requirements (the maximum input voltage of 0.5V must be able to be amplified to 20V), while the resistances are again educated guesses.

Now consider the power output stage.

Module	Power Output Stage
Inputs	- Audio input signal: <u>20</u>V peak.
	- Power: ± <u>25</u>V DC.
Outputs	- Audio signal: <u>20</u>V peak at up to <u>2.5</u>A.
Functionality	Provide unity voltage gain with output current as required by a resistive load of up to <u>2.5</u>A. It should have an input resistance >$\underline{1\text{M}\Omega}$ and output resistance <<u>1</u>Ω.

For the power output stage, it is clear that 20V peak needs to be delivered, but how was the requirement on current determined? The current needed to drive the speaker is determined from Ohm's Law as $I = V/R = 20/8 = 2.5A$.

The last module to examine at this level is the power supply.

Module	Power Supply
Inputs	- 120 Volts AC rms.
Outputs	- Power: ± <u>25</u>V DC with up to <u>3.0</u> A of current with a regulation of <<u>1</u>%.
Functionality	Convert AC wall outlet voltage to positive and negative DC output voltages, and provide enough current to drive all amplifiers.

It is clear that the power supply needs to deliver ± 25V DC, while the 3.0A current capability was selected to supply the 2.5A needed for the peak output power requirement plus the current needed to power the other amplifier stages.

Finally, it is necessary to determine if the values of the input and output resistances selected for the stages are realistic. For cascaded amplifier stages, the overall voltage gain is given by the product of gains multiplied by the voltage divider losses between stages [Sed04]. In this case the overall gain is

$$\text{voltage gain} = \text{gain}_1 \times \text{gain}_2 \times \text{gain}_3 \left(\frac{R_{in2}}{R_{in2} + R_{out1}} \right) \left(\frac{R_{in3}}{R_{in3} + R_{out2}} \right)$$

$$= 1 \times 40 \times 1 \left(\frac{100\text{k}}{100\text{k} + 100} \right) \left(\frac{1\text{M}}{1\text{M} + 100} \right) \qquad (1)$$

$$\approx 40$$

So the resistance values selected in the functional requirements satisfy the overall system requirements. If not, it would be necessary to go back and refine them.

Level 2

At this point, the three amplifier stages are ready for detailed component level design, while the power supply needs another level of refinement as shown in Figure 5.4. The functional requirement for each of the elements in the power supply would be developed similarly. Functional decomposition stops at this point—all levels of the hierarchy are defined and the next step is the detailed design, where the actual circuit components are determined.

Figure 5.4 Level 2 design of the power supply.

5.5 Application: Digital Design

Functional decomposition is widely applied to the design of digital systems, where it is known as *entity-architecture* design. The inputs and outputs refer to the entity, and the architecture describes the functionality. The application of functional decomposition to digital systems is demonstrated in the following example. Consider the design of a simple digital stopwatch that keeps track of seconds and has the following engineering requirements.

The system must

- Have no more than two control buttons.
- Implement Run, Stop, and Reset functions.
- Output a 16-bit binary number that represents seconds elapsed.

Level 0

The Level 0 diagram and functional requirements are shown below.

Figure 5.5 Level 0 digital stopwatch functionality.

Module	Stopwatch
Inputs	- A: Reset button signal. When the button is pushed, it provides a logic level high signal (5V) that resets the counter to zero. - B: Run/stop toggle signal. When the button is pushed, it provides a logic level high signal (5V) that toggles between run and stop modes.
Outputs	- b_{15}-b_0: 16-bit binary number that represents the number of seconds elapsed.
Functionality	The stopwatch counts the number of seconds after B is pushed when the system is in the Reset or Stop mode. When in Run mode and B is pushed, the stopwatch stops counting. A reset button push (A) will reset the output value of the counter to zero only when in Stop mode.

Level 1

The Level 1 architecture in Figure 5.6 contains three modules: a seconds counter, a clock divider, and a finite state machine (FSM). The stopwatch counts seconds, thus the seconds counter module counts the seconds and outputs a 16-bit number representing the number of seconds elapsed. The clock divider generates a 1Hz signal that triggers the seconds counter. The FSM responds to the button press stimuli and produces the appropriate control signals for the seconds counter. The system clock is included to clock both the FSM and the clock divider.

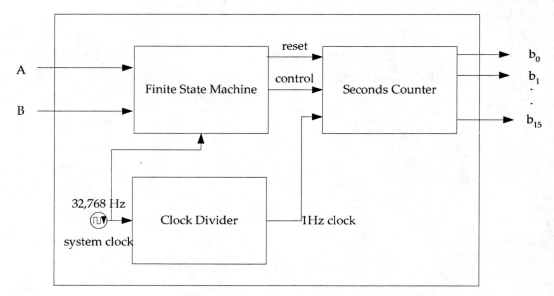

Figure 5.6 Level 1 design for the digital stopwatch.

The functionality of the Level 1 modules is described as follows, starting with the finite state machine.

Module	Finite State Machine
Inputs	- A: Signal to reset the counter. - B: Signal to toggle the stopwatch between run and stop modes. - Clock: 1Hz clock signal.
Outputs	- Reset: Signal to reset the counter to zero. - Control: Signal that enables or disables the counter.
Functionality	

The functionality of the finite state machine is described with a tool that is probably familiar to the reader, the state diagram. State diagrams are covered in more detail in Chapter 6. The state diagram describes stimulus-response behavior, and shows how the system transitions between states based upon logic signals from the button presses.

Next, consider the clock divider.

Module	Clock Divider
Inputs	- System clock: 32,768Hz.
Outputs	- Internal clock: 1Hz clock for seconds elapsed.
Functionality	Divide the system clock by 32,768 to produce a 1Hz clock.

The value of 32,768 Hz was selected for the system clock for several reasons. It is a power of 2 that is easily divisible by digital circuitry to produce a 1Hz output signal. It is also well above the clock rate needed for detecting button presses and there is a wide selection of crystals that can meet this requirement.

Finally, consider the seconds counter.

Module	Seconds Counter
Inputs	- Reset: Reset the counter to zero. - Control: Enable/disable the counter. - Clock: Increment the counter.
Outputs	- b_{15}-b_0: 16-bit binary representation of number of seconds elapsed.
Functionality	Count the seconds when enabled and resets to zero when reset signal enabled.

The system decomposition would end here, assuming that the design is to be implemented using off-the-shelf chips. The next step would be to determine components at the detailed design level. However, if it were an integrated circuit design, the description would continue until the transistor level is reached.

5.6 Application: Software Design

Software also lends itself to functional decomposition since virtually all computing languages provide the capability to call functions, subroutines, or modules. Functional software design simplifies program development by eliminating the need to create redundant code via the use of functions that are called repeatedly.

Structure charts are specialized block diagrams for visualizing functional software designs. The modules used in a structure chart are shown in Figure 5.7. The larger arrows indicate connections to other modules, while the smaller arrows represent data and control information passed between modules. Five basic modules are utilized:

1. *Input modules.* Receive information.
2. *Output modules.* Return information.
3. *Transform modules.* Receive information, change it, and return the changed information.
4. *Coordination modules.* Coordinate or synchronize activities between modules.
5. *Composite modules.* Any possible combination of the other four.

This approach to software design, also known as *structured design*, was formalized in the 1970's by IBM researchers [Ste99].

Figure 5.7 Module types for functional software design. The larger arrows indicate connections between modules and the smaller arrows represent data and control.

The following example demonstrates the application of functional decomposition to a software design with the following requirements.

The system must

- Accept an ASCII file of integer numbers as input.
- Sort the numbers into ascending order and save the sorted numbers to disk.
- Compute the mean of the numbers.
- Display the mean on the screen.

This is a fairly simple task that could easily be done in a single function, but doing so would not allow components of the design to be easily reused, tested, or troubleshot. The engineering requirements themselves provide some guidance in terms of how to arrange the functionality of the modules (*form follows function*). The architecture in Figure 5.8 contains a `main` module that calls three sub-modules. In this design `main` is a coordinating module that controls the processing and calling of the other modules, a common scenario. It was also decided that all user interaction would take place within `main`. The order of the processing is not described by structure charts. In our program, `main` calls `ReadArray`, `SortArray`, and `ComputeMean` in sequential order. `main` passes the filename (`fname`) to `ReadArray`, which reads in the array and the number of elements in it, and returns this information to `main`. The choice of passing in the filename was deliberate; the user could have been prompted for the filename in `ReadArray`, but doing so might limit future reuse of the function since you may not always want to do so when reading an array of data. `SortArray` is then called, which accepts the array of numbers and the number of elements in the array, and returns the sorted values in the same array. Finally, `ComputeMean` is executed, which accepts the sorted array and the number of elements, computes the mean value, and returns it to `main`.

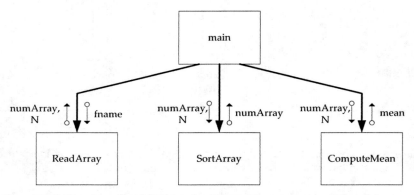

Figure 5.8 Structure chart design of sorting and mean computation program.

The functional requirements for each module in the structure chart are detailed in Table 5.1. The structure chart provides a visual relationship between modules in the design, but also has some disadvantages. It is difficult to visualize designs as the complexity of the software increases. This can be addressed by expanding sublevels in the design as necessary in different diagrams. Structure charts also lack a temporal aspect that indicates the calling order. Most software systems have many layers in the hierarchy and highly complex calling patterns. In this example, `main` calls three modules in a well-defined order, but if there was another level in the hierarchy, there is no reason why it could not be called by a module at any other level. That leads to some of the unique problems associated with software design. Functional design works well for small-to-moderately complex software, but tends to fall short when applied to large scale software systems. As such, it has given way to the object-oriented design approach.

5.7 Application: Thermometer Design

The final example includes both analog and digital modules where the objective is to design a thermometer that meets the following engineering requirements.

The system must

- Measure temperature between 0 and 200°C.

- Have an accuracy of 0.4% of full scale.

- Display the temperature digitally, including one digit beyond the decimal point.

- Be powered by a standard 120V 60Hz AC outlet.

- Use an RTD (thermal resistive device) that has an accuracy of 0.55°C over the range. The resistance of the RTD varies linearly with temperature from 100Ω at 0°C to 178Ω at 200°C. (Note: this requirement does not meet the abstractness property identified in Chapter 3, since it identifies part of the solution. This requirement is given to provide guidance in this example.)

Table 5.1 Functional design requirements for the number sort program.

Module name	main()
Module type	Coordination
Input arguments	None
Output arguments	None
Description	The main function calls ReadArray() to read the input file from disk, SortArray() to sort the array, and ComputeMean() to determine the mean value of elements in the array. User interaction requires the user to enter the filename, and the mean value is displayed on the screen.
Modules invoked	ReadArray, SortArray, and ComputeMean

Module name	ReadArray()
Module type	Input and output
Input arguments	- fname[]: character array with filename to read from.
Output Arguments	- numArray[]: integer array with elements read from file. - N: number of elements in numArray[].
Description	Read data from input data file and store elements in array numArray[]. The number of elements read is placed in N.
Modules invoked	None

Module name	SortArray()
Module type	Transformation
Input arguments	- numArray[]: integer array of numbers. - N: number of elements in numArray[].
Output Arguments	- numArray[]: sorted array of integer numbers.
Description	Sort elements in array using a shell sort algorithm. Saves the the sorted array to disk.
Methods invoked	None

Module name	ComputeMean()
Module type	Input and output
Input arguments	- numArray[]: integer array of numbers. - N: number of elements in numArray[].
Output arguments	- mean: mean value of the elements in the array.
Description	Computes the mean value of the integer elements in the array.
Modules invoked	None

Level 0

The overall goal is to convert a sensed temperature to a digital temperature reading. The Level 0 description is as follows.

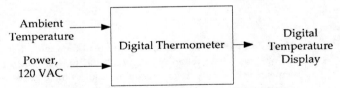

Figure 5.9 Level 0 digital thermometer functionality.

Module	Digital Thermometer
Inputs	- Ambient temperature: 0-200°C. - Power: 120V AC power.
Outputs	- Digital temperature display: A four digit display, including one digit beyond the decimal point.
Functionality	Displays temperature on digital readout with an accuracy of 0.4% of full scale.

Level 1

The Level 1 architecture selected is shown in Figure 5.10. The temperature conversion unit converts the temperature to an analog voltage using the RTD that is sampled by the analog-to-digital converter. The N-bit binary output from the converter is translated into binary-coded decimal (BCD). BCD is a 4-bit representation of the digits between 0 and 9. Since there are four display digits, there are four separate binary encoded outputs from the BCD conversion unit. Common 7-segment LEDs are used for the display. However, they do not directly accept BCD and instead have seven input lines, each of which is individually switched to control the display segments. The requirements did not specifically address cost or size constraints, nor clearly define the environment, so there are many possible solutions. For example, an analog-to-digital current converter could be used, integrated circuit temperature sensing packages could be considered, and microcontroller-based solutions are feasible as well.

From a system design perspective, an error budget is needed to identify the maximum error that each subsystem may introduce, while still achieving the overall accuracy. In this case, error is introduced in the temperature conversion unit and A/D converter, but not in the remaining digital components. The overall accuracy that the system must achieve is 0.4%, and that translates into 0.8°C of allowable error for the 200°C range. Let's now examine the modules.

Figure 5.10 Level 1 design of the digital thermometer.

The functionality of the Level 1 modules is described as follows, starting with the temperature conversion unit.

Module	Temperature Conversion Unit
Inputs	- Ambient temperature: 0-200°C. - Power: ?V DC (to power the electronics).
Outputs	- V_T: temperature proportional voltage. $V_T = \alpha T$, and ranges from ? to ?V.
Functionality	Produces an output voltage that is linearly proportional to temperature. It must achieve an accuracy of ?%.

There are several unknowns at this point. The voltage necessary to power the electronics is not known, but a reasonable assumption could be made. The output voltage range and the accuracy are unknown. It is known that the RTD will introduce up to 0.55°C of error and that the electronics themselves will introduce additional error (the exact amount is unknown at this point). An educated guess is made that the maximum error allowed for the temperature unit is 0.6°C. This means that the electronics themselves would be required to introduce no more than 0.05°C of error due to the 0.55°C of error introduced by the RTD.

Now consider the analog to digital converter.

Module	A/D Converter
Inputs	- V_T: voltage proportional to temperature that ranges from ? to ?V. - Power: ?V DC.
Outputs	- $b_{N-1} - b_0$: ?-bit binary representation of V_T.
Functionality	Converts analog input to binary digital output.

The A/D converter is not likely to be something that is designed due to the availability of low cost, off-the-shelf solutions. The requirements drive the converter selection. There are two unknowns—the number of bits and the range of the input voltage. The number of bits affects the accuracy, since the greater the number of bits, the better the accuracy. The number of bits needed for the converter is calculated from the maximum allowable error that the A/D can introduce (0.2°C), the number of discrete intervals, and the temperate range as

$$\text{max error} = \frac{\text{range}}{\text{number of intervals}} = \frac{200°C}{2^N} \leq 0.2°C \quad \Rightarrow N \geq 9.97 \text{ bits}. \tag{2}$$

So the A/D converter needs to have at least 10 bits. How is the voltage range selected? It is typically fixed for a particular integrated circuit solution, but the temperature conversion subsystem output should be matched to the voltage range so that all bits are effectively utilized, otherwise, error is introduced.

Now, consider the BCD conversion unit.

Module	BCD Conversion Unit
Inputs	- 10-bit binary number (b_9-b_0): Represents the range 0.0-200.0°C. - Power: ?V DC.
Outputs	- BCD_0: 4-bit BCD representation of tenths digit (after decimal). - BCD_1: 4-bit BCD representation of one's digit. - BCD_2: 4-bit BCD representation of ten's digit. - BCD_3: 4-bit BCD representation of hundred's digit.
Functionality	Converts the 10-bit binary number to BCD representation of temperature. Must refresh the displays twice a second.

The objective of the BCD conversion unit is fairly simple, although the component level design of the circuitry to accomplish the conversion is not.

This leads to the last module, the 7-segment LED driver, whose functionality is described as follows.

Module	7-Segment LED Driver
Inputs	- BCD_0: 4-bit BCD representation of tenths digit (after decimal). - BCD_1: 4-bit BCD representation of one's digit. - BCD_2: 4-bit BCD representation of ten's digit. - BCD_3: 4-bit BCD representation of hundred's digit. - Power: ?V DC.
Outputs	- Four 7-segment driver lines.
Functionality	Converts the BCD for each digit into outputs that turn on LEDs in 7-segment package to display the temperature.

For completeness, the functional requirements of the power supply are supplied. They are similar to the power supply requirements utilized in the audio amplifier design in Section 5.4.

Module	Power supply
Inputs	- 120 Volts AC rms.
Outputs	- ± ?V DC with up to ?mA of current.
	- Regulation of ?%.
Functionality	Convert AC wall outlet voltage to positive and negative DC output voltages, with enough current to drive all circuit subsystems.

At this point, the requirements for the major subsystems are completed and ready for design at the component level. Illustration of the complete design would require a fair amount of detail, and while it is not presented here, some of the issues involved are discussed. First, there are a variety of electronic circuits (inverting op amps, single BJT configurations, and current mirrors, etc.—see Example 4.1 in Chapter 4) that could be utilized as a current source to drive the RTD in the temperature conversion subsystem. A midrange resolution A/D converter is needed, and its particular input voltage range drives the output voltage requirements for the temperature conversion module. The BCD conversion circuitry could be implemented using combinational digital logic (tedious due to the number of discrete gates), or a more efficient, but slower, sequential logic design. Finally, the 7-segment display converters could be designed using combinational logic that maps the BCD inputs into outputs to activate the appropriate display segments.

5.8 Coupling and Cohesion

The concepts of coupling and cohesion are examined before concluding this chapter. They originated to describe software designs [Ste99], but are applicable to electrical and computer systems. To understand their importance, consider the relationship between the number of modules in a system and the number of connections between them. For our purposes, a connection between two modules may consist of any number of signals without regard to their direction. Thus, a system consisting of two modules has, at most, one connection. If the number of modules is increased to three, the number of possible connections increases to three, a system with four modules has six possible connections, and five modules increases the number of possible connections to ten. The point is that the maximum number of potential connections increases rapidly with the number of modules in the system. The relationship between the maximum possible connections and number of modules (n) is given by

$$\text{connections}_{max} = \frac{n(n-1)}{2} . \tag{3}$$

Modules are coupled if they depend upon each other in some way to operate properly. *Coupling* is the extent to which modules or subsystems are connected [Jal97]. Although there is no agreed upon mathematical definition of coupling, it seems obvious that increasing the exchange of control and data between two modules leads to a higher degree of coupling. When systems are highly coupled, it is difficult to change one module without impacting the other. Consider the extreme case where all modules in a system are connected to each other—an error in one module has the potential to impact every other module in the system. Errors in a module are propagated to others to a degree that is related to the amount of coupling. From this point of view, it is good to minimize coupling. Yet coupling cannot be eliminated, since the point of functional decomposition is to break a design into components that work together to produce a higher level behavior.

There are two ways to reduce coupling—minimize the number of connections between modules and maximize cohesion within modules. *Cohesion* refers to how focused a module is—highly cohesive systems do one or a few things very well. Stevens *et al.* [Ste99] defined six types of cohesion from the weakest to strongest as: coincidental, logical, temporal, communicational, sequential, and functional. More information on this can be found in the original work, but the conclusion is that modules with high functional cohesion are the most desirable. So it is best to design modules with a single well-defined functional objective consistent with the philosophy of functional decomposition. This leads to the important design principle that it is desirable to maximize cohesion, while minimizing coupling.

Coupling and cohesion impact the later stages of testing and system integration. If a particular module is highly cohesive, then it should be possible to test it independently of the other modules to verify its operability. This does not mean that it will necessarily operate properly when integrated into the overall system, but the probability that it will is higher if provided with proper inputs from connected modules. Contrast this to the case of a low cohesion system. In that case, it will likely be difficult to test the individual modules without first integrating them.

To develop a better understanding, consider the amplifier design in Figure 5.3 (Section 5.4) with three cascaded amplifier stages. Each stage is highly cohesive, performing a singular function of signal amplification. Each of these stages could easily operate as a stand-alone module independent of the complete system. How about coupling? In terms of the number of connections, it is fairly low as each amplifier stage has an input and output voltage signal. The most coupled module in the system is the power supply, and not surprisingly, its failure leads to a complete system failure. Coupling in this case can also be viewed in terms of the resistance matching between input and output of the cascaded stages, producing the voltage divider effect in equation (1). For voltage amplifiers, the goal is to have high input resistance and low output resistance, which minimize both voltage losses and coupling. The stages are not completely uncoupled, because the input resistances, although large, are not infinite, and the output resistances are not zero. The modules in the power supply unit in Figure 5.4 (rectifier, smoothing filter, and regulator) have a much higher degree of coupling. In fact, it is difficult to develop a clear functional decomposition of the power supply module because the elements in the smoothing filter also serve as part of the rectifier circuit (refer to a basic electronics textbook [Sed04] for more information).

As another example, consider a software design where two options are under consideration: one large function with 1000 lines of code, versus 15 cohesive functions, each with an average of 100 lines of code. Both perform the same function, but which runs faster? Most likely the first, as it would be highly integrated and would not suffer from overhead needed with multiple functions. Which is easier to upgrade and debug a year from now? That is clearly the second case. Although loosely coupled and highly cohesive designs may facilitate better design and testing, they may not be best in terms of performance.

5.9 Project Application: The Functional Design

The following is a format for documenting and presenting functional designs.

Design Level 0

- Present a single module block diagram with inputs and outputs identified.
- Present the functional requirements: inputs, outputs, and functionality.

Design Level 1

- Present the Level 1 diagram (system architecture) with all modules and interconnections shown.
- Describe the theory of operation. This should explain how the modules work together to achieve the functional objectives.
- Present the functional requirements for each module at this level.

Design Level N (for N>1)

- Repeat the process from Design Level 1 for as many levels as necessary.

Design Alternatives

- Describe the different alternatives that were considered, the tradeoffs, and the rationale for the choices made. This should be based upon concept evaluation methods communicated in Chapter 4.

5.10 Summary and Further Reading

This chapter presented the functional decomposition design technique, where every level of the design is decomposed into sub-modules, each of which is the domain of the next lower level. The inputs, outputs, and functionality must be determined for a given module. Applying the process in Figure 5.1 and following the guidelines in Section 5.3 should aid in the application of functional decomposition. Functional decomposition is applicable to a wide variety of systems, and in this chapter designs of analog electronics, digital electronics, and software were examined.

Nigel Cross presents a good overview of the functional decomposition method with application to mechanical systems [Cro00], but with less focus on the description of the functional requirements than presented here. The work by Stevens *et al.* [Ste99] is interesting reading that gives an understanding of the evolution of structured design. It delves into the concepts of coupling and cohesion. Coupling and cohesion are also addressed well in the book by Jalote [Jal97]. An in-depth treatment of structured systems design is found in <u>The Practical Guide to Structured Systems Design</u> [Pag88]. This guide also integrates data flow diagrams with functional techniques. Finally, the thermometer design example was inspired by Stadtmiller's book [Sta01] on electronics design.

5.11 Problems

5.1   Describe the differences between *bottom-up* and *top-down* design.

5.2   Develop a functional design for an audio graphic equalizer. A graphic equalizer decomposes an audio signal into component frequencies bands, allows the user to apply amplification to each individual band, and recombines the component signals. The design can employ either analog or digital processing. Be sure to clearly identify the design levels, functional requirements, and theory of operation for the different levels in the architecture.

The system must

- Accept an audio input signal source, with a source resistance of 1000Ω and a maximum input voltage of 1V peak-to-peak.
- Have an adjustable volume control.
- Deliver a maximum of 40W to an 8Ω speaker.
- Have four frequency bands into which the audio is decomposed (you select the frequency ranges).

5.3   Develop a functional design for a system that measures and displays the speed of a bicycle. Be sure to clearly identify the design levels, functional requirements, and theory of operation for each level.

The system must

- Measure instantaneous velocities between zero and 75 miles per hour with an accuracy of 1% of full scale.

- Display the velocity digitally and include one digit beyond the decimal point.

- Operate with bicycle tires that have 19, 24, 26, and 27 inch radii.

5.4 Draw a structure chart for the following C++ program:

```cpp
void IncBy5(int &a, int &b);
int  Multiply(int a, int b);
void Print(int a, int b);

main(){
    int x=y=z=0;
    IncBy5(x,y);
    z=Mult(x,y);
    Print(x,z);
}
void IncBy5(int &a, int &b) {
    a+=5;
    b+=5;
    Print(a,b);
}

int Multiply(int a, int b){
  return (a*b);
}

void Print(int a, int b) {
    cout << a << ", " << b;
}
```

5.5 Develop a functional design for software that meets the following requirements.

The system must

- Read an array of floating point numbers from an ASCII file on disk.
- Compute the average, median, and standard deviation of the numbers.
- Store the average, median, and standard deviation values on disk.

The design should have multiple modules and include the following elements: (a) a structure chart, and (b) a functional description of each module.

5.6 Describe in your own words what is meant by coupling in design. Describe the advantages of both loosely and tightly coupled designs.

5.7 **Project Application.** Develop a functional design for your project. Follow the presentation guidelines in Section 5.9 for communicating the results of the design.

Chapter 6 System Design II: Behavior Models

Genius is 1% inspiration and 99% perspiration. —Thomas Edison

The functional decomposition technique examined in Chapter 5 is a powerful modeling tool for system design that is applicable for describing functional behavior (input, output, transform). However, that approach by itself is limited in its descriptive ability. This was apparent in the digital stopwatch example that required the use of state diagrams, in addition to functional decomposition, to fully articulate the design. State diagrams represent stimulus-response behavior for systems with memory. This means that they describe how systems transition between states based upon stimuli and knowledge of past states. Models provide a convenient method of describing a system at a high level of abstraction. This allows systems to be described without having to determine all of the implementation details. All models are not the same—they come in a variety of forms and each serves a different intention.

   The objective of this chapter is to provide an overview of other design tools for describing system behavior, with an emphasis on computing systems. The first tool examined is the state diagram. This is followed by the flowchart, which is used to describe algorithmic processes and logical behavior. Two modeling languages for describing information and data handling—data flow diagrams and entity relationship diagrams—are then examined. The final topic considered is the Unified Modeling Language, which is a collection of system views that are used to describe behavior.

Learning Objectives

By the end of this chapter, the reader should:

- Have a familiarity with the following modeling tools for describing ECE system behavior: state diagrams, flowcharts, data flow diagrams, entity relationship diagrams, and the Unified Modeling Language.
- Understand the intention and expressive power of the different models.
- Understand the domains in which the models apply.
- Be able to conduct analysis and design with the models.
- Understand what model types to choose for a given design problem.

6.1 Models

From the previous chapter we know that the top-down design process starts with an abstraction of the system to be built. This initial design is called an abstraction because it is not an actual physical realization. An abstraction that is expressed in a standardized and accepted language is called a model. A modeling language does not have to be formed from letters and words—often the words are graphical symbols. You are already familiar with many different models from everyday life such as blueprints, a diagram of a football play, knitting instructions, electrical schematics, and mathematical formulas to name a few. In order to be effective, a model should meet the following properties [Sat02]:

- *Be abstract.* This means that the model should be independent of final implementation and that there should be multiple ways of implementing the design based upon it.

- *Be unambiguous.* A model should have a single clear meaning in terms of describing the intended behavior.

- *Allow for innovation.* Models should encourage exploration of alternative system implementations and behaviors.

- *Be standardized.* Standardization provides a common language that can be understood by all. Designers should be wary of developing their own models that are ill-defined and not commonly understood.

- *Provide a means for communication.* A model should facilitate communication within the design team and with non-technical stakeholders.

- *Be modifiable.* A model should make design modifications relatively easy.

- *Remove unnecessary details and emphasize important features.* The intent is to simplify the design for ease of understanding. The most highly detailed information is typically identified in the detailed design.

- *Break the overall problem into sub-problems.* Most problems are too complex to be handled directly and must be decomposed into subsystems. This produces the design hierarchy.

- *Substitute a sequence of actions by a single action.* This allows understanding of the overall larger behavior, which can then be examined at other levels. This supports the ability to break a design into sub-problems.

- *Assist in verification.* A model should aid in demonstrating that the design meets the engineering requirements.

- *Assist in validation.* Validation is the process of demonstrating that the needs of the user are being met and the right system is being designed. The model should facilitate discussion with all stakeholders to ensure it meets everyone's expectations.

These properties are similar to those identified in Chapter 3 for both an engineering requirement and the overall Requirements Specification. As such, models are often referred to as the *Design Specification* since they represent a specification of the system behavior at a more detailed level. In order to meet these properties, most models have an ***object type***, which is capable of encapsulating the actual components used to construct the target system. A model also has something known as an ***intention,*** which is the behavior that it describes. For example, compare the intention of a circuit schematic to the schematic of a football play. Since models are built with different intentions, it possible to choose the wrong model for a particular system—it would surprise a football team to see a play schematic represented with resistors and capacitors!

6.2 State Diagrams

State diagrams are used to describe the behavior of systems with memory. A system with memory is able to modify its response to inputs based on the state of the system. The ***state*** of a system represents the net effect of all previous inputs to the system. Since the state characterizes the history of previous inputs, it is often synonymous with the word memory. Intuitively, a state corresponds to an operating mode of a system, and inputs are associated with transitions between states. To determine if a system has memory, ask the following question— *"Can the same input produce different outputs?"* If the answer to this question is yes, then the system has memory and can be modeled with a state diagram.

A state diagram is a drawing that consists of states and transition arcs as shown in Figure 6.1. Each state is represented as a rectangle with rounded edges with the name of the state written inside. Whenever possible, states should be given meaningful names. When there exists the possibility for ambiguity in the names, a table should be created that identifies the state names and their associated meanings. There are special circle symbols for both initial and final states. Transitions are drawn as arrows from a source state to a destination state. Since inputs cause the transitions between states, the arrows are labeled with their associated inputs. The outputs are listed directly in the state since it is assumed that the outputs are associated with states.

Figure 6.1 Symbols used in state diagrams.

As an example, consider a vending machine that accepts nickels and dimes and dispenses a piece of candy when 25 cents has been deposited. This vending machine can be modeled using a state diagram because the response of the machine to a coin depends on how much money has been deposited so far. One possible state diagram for the vending machine is shown in Figure 6.2.

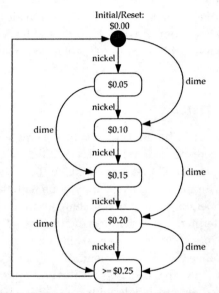

Figure 6.2 A state diagram for a simple vending machine.

The initialization/reset state is labeled $0.00, while the action of the machine (dispense or not dispense candy) is associated with the states—the only state that dispenses candy is $0.25. There are three types of transitions shown in this state diagram. The two labeled nickel and dime are associated with depositing those coins respectively. The unlabeled transition from state $0.25 to state $0.00 is called an unconditional transition. A system with an unconditional transition between two states is assumed to remain in the first state for a defined time period before moving to the second state. In this case it ensures that the machine dispenses a single candy before going on to the next transaction. It is common practice in state diagrams to assume that any unspecified input conditions cause the system to remain in the current state. For example, if no coin is inserted while in state $0.20 the system remains in state $0.20. Finally, note that since this machine does not dispense change, it can overcharge a customer for candy. This would not be a popular vending machine with users!

The vending machine is a fairly traditional application of a state diagram. However, state diagrams can describe the process of a system to non-experts in a very intuitive way. For example, consider the audio record and playback unit described by the state diagram shown in Figure 6.3.

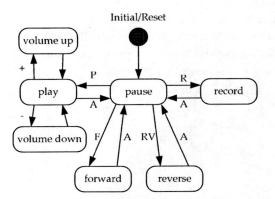

Stimulus	Description
P	Play button pressed.
R	Record button pressed.
F	Forward button pressed.
RV	Reverse button pressed.
A	Any (P,R,F,RV) button pressed.
+	Increase volume button.
-	Decrease volume button.

Figure 6.3 A state diagram for an audio record and playback unit.

The system automatically starts in the initialization state and moves directly into the pause state, where it waits for a command. When the record button 'R' is pressed, the system starts recording, and if any button (excluding the volume button) is pressed, it returns to the pause state. The operation is similar for the forward and reverse states. When in the play state, it is also capable of changing the system volume if either the increase or decrease volume buttons is pressed. Each of these would change the volume by a set amount and then return automatically to the play state.

The description of the audio unit lacks many important details. For example, the record state may actually consist of a sequence of states needed to initialize the recording media, transfer data to the media, and check for a button press. However, this abstracting out the details of a system allows the formal behavior of a system to be described to a non-technical audience.

6.3 Flowcharts

The intention of a *flowchart* is to visually describe a process or algorithm, including its steps and control. Flowcharts are often scoffed at as being old-fashioned and overly simple—these criticisms are actually strengths. Since flowcharts have been around for a long time, they are easily recognized and understood. Furthermore, being simple means that learning how to create and understand them is relatively easy. Due to their simplicity, they are applied in a great number of applications, including non-technical ones such as the description of business processes.

Some of the primary symbols used in flowcharts are shown in Figure 6.4. The names of the starting and ending steps of a flowchart are represented by ovals known as terminators. Individual processing steps are written inside of rectangles, while a process step that is elaborated by another flowchart is drawn as a rectangle with double sides. Elaborated processes allow the representation of hierarchy in the design. Certain points in a flowchart can lead to alternative destinations as represented by a decision or conditional symbol (diamond). The condition that determines the next step is written inside the diamond and the possible values of the condition are written on the arcs leaving the conditional step. As shown in Figure 6.4 there are multiple ways to indicate data stores for retrieving or saving data.

Figure 6.4 Basic flowchart symbols.

As an application, consider an embedded computer system that monitors the light level of its environment as described by the flowchart in Figure 6.5. The algorithmic process of the flowchart is easy and intuitive to understand. The system reads the current light value, stores it into an array, and then computes an average. In a complete system description, there would be a second flowchart describing how the system determines the average of the values of the light samples, since this is identified as an elaborated process. The system then waits 1ms, writes the average to a terminal (display device), and then checks for a key-press. Light levels continue to be monitored in the absence of a key-press, otherwise the light monitoring process halts.

Figure 6.5 A flowchart for an embedded system.

Flowcharts are an intuitive way to describe algorithmic processes. Limiting the complexity of a flowchart to between 10 and 20 steps enables the sequence of actions to be quickly comprehended while eliminating unnecessary details. Flowcharts are effective for describing common algorithmic control and decision structures. Examples of the commonly used do-while, do-until, and nested control structures are shown in Figure 6.6. What flowcharts do not represent is the structure of the data being manipulated and they are not particularly good at representing concurrent processes.

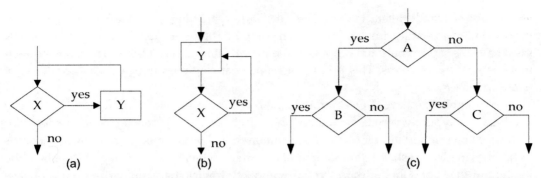

Figure 6.6 Flowchart control and decision structures. (a) do Y while X is true. (b) do Y until X is not true. (c) Nested decision structure.

6.4 Data Flow Diagrams

The intention of a *data flow diagram* (DFD) is to model the processing and flow of data inside a system. It is a function-oriented approach that is similar to functional decomposition—the processes inside a DFD accept data inputs, transform them in some way, and produce output data. A DFD is often used for the analysis of information systems due to its data emphasis, but can be broadly applied to ECE systems. It differs from functional decomposition in that functional decomposition is often closer to the implementation of the design, whereas the DFD models the system from a data point of view. A DFD is fundamentally different from a flowchart in that it does not encapsulate control and sequencing information, but allows multiple processes running concurrently. There are four symbols, shown in Figure 6.7, that are used in a DFD:

1. *Processes.* A rectangle with rounded corners that describes a useful task or function. They perform a transformation on the data.
2. *Data flows.* An arrow representing a data relationship between two processes.
3. *Data stores.* An open rectangle representing a data repository.
4. *Interfaces.* A square describing external agents or entities that use the system. They are also referred to as sources and sinks.

Figure 6.7 Data flow diagram symbols.

Like the general design process, DFDs are successively refined from the top-down. That is, there is a single top level (or Level 0) DFD describing the entire system and the interfaces and data stores that it interacts with. The rules for constructing a DFD are fairly intuitive. A process must have at least one input and one output. The refinement of a process at level N must have the same inputs and outputs as the process at level N-1. Data must be transformed in some way by a process. This process of refinement continues until a satisfactory level of detail is reached.

An example of a Level 1 DFD for a video browsing system is shown in Figure 6.8. Video databases are typically very large; due to their size, it is usually cumbersome to preview videos and extract important information. A solution to this problem is to apply image analysis techniques to identify shots (continuous scenes without a break) in a video and store both the location of the shot boundaries in the video and key frames that summarize each shot. The collection of key frames is known as a storyboard. The storyboards are stored in an annotation database that is much smaller in size than the original video database. Typically the user of a video browsing system has the ability to preview the storyboard and select shots from it to view. This example shows the data flow for such a system.

The data processing is readily apparent from the DFD. Videos in the database are processed to extract shot boundaries and key frames, which are stored in the annotation database. The user can submit requests to view a storyboard, which is retrieved from the annotation database. When a shot is selected from the storyboard, the users can preview the original shot, which is retrieved from the video database.

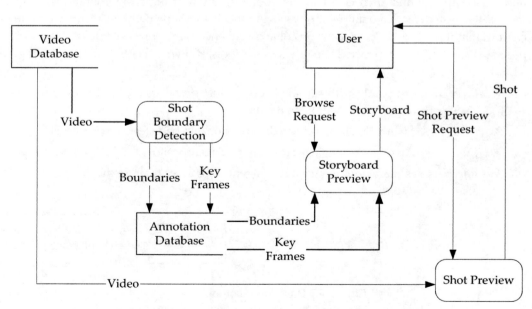

Figure 6.8 Level 1 data flow diagram for a video browsing system.

There are a few important points to note about DFDs. They are solution independent, specifying the behavior based upon data flow. Specific information on the data flows is defined in a formal language known as a *data dictionary* (it is not covered here). There can be concurrent processes represented in a DFD. In the video browsing example, multiple people can use the system simultaneously, and there is no implied sequencing between when the shot boundary detection process is run and the storyboards are previewed. It is clear that the video must undergo shot detection prior to viewing its storyboard. This information is listed in something known as an *event table*. The event table for this example is shown in Table 6.1.

Table 6.1 Event table for the video browsing system.

Event	Trigger	Process	Source
Annotate Video	New Video Arrival	Shot Boundary Detection	System
View Storyboard	Browse Request	Storyboard Preview	User
View Shot	Shot Preview Request	Shot Preview	User

An *event* is an occurrence at a specific time and place that needs to be remembered. Events can be classified into temporal, external, and state. A *temporal event* is one which happens because the system has reached some critical time. In this example, the generation of shot boundaries could occur for all new videos in the database at a specified time each day, but in this case it occurs whenever a new video is added to the database. An *external event* occurs outside the system boundary by a system user, in this case requesting either a storyboard or a video preview. A *state event* is the result of something changing within the system. Associated with each event is a *trigger*, the cause of the event. Each event has a process that is associated with it. Finally, associated with each event is a *source*, the entity responsible for triggering it.

6.5 Entity Relationship Diagrams

A database is a system that stores and retrieves data, and it is modeled by an *entity relationship diagram* (ERD). The intention of an ERD is to catalog a set of related objects (entities), their attributes, and the relationships between them. The entities and their relationships are real distinct things that have characteristics that need to be captured. The design of a database starts by describing the entities, their attributes, and the relationship between entities in an ERD. In order to ask meaningful questions about the data, the entities need to be related to one another. For example, a list of students and a list of courses by themselves have limited utility. However, by introducing a relationship between these two entities we can ask questions such as "*How many students are taking the Microelectronics course?*" The three elements used in the ERD modeling language are:

1. *Entities.* They are generally in the form of tangible objects, roles played, organizational units, devices, and locations. An instance is the manifestation of a particular entity. For example, an entity could be Student while an instance would be Kristen.
2. *Relationships.* They are descriptors for the relationships between entities.
3. *Attributes.* They are features that are used to differentiate between instances of the entities.

ERDs are examined through the example of a database system for a college. The process typically starts by interviewing the end-users and identifying the entities and their attributes. Assume that the result of this process is that the college wants to store data about three entities: Students, Courses, and Departments. The process of building an ERD starts by determining the relationships between entities. One way to do this is to build an *entity relationship matrix* as shown in Table 6.2. The entities constitute both the row and column headings, and the matrix entries represent the relationship, if any, that exists between entities. This is similar to the pairwise comparison matrix in Chapter 2 where user needs were systematically compared. Relationships are bi-directional because they have two participating entities.

Table 6.2 Entity relationship matrix.

	Student	**Course**	**Department**
Student		takes many	majors in one
Course	has many	can require many / can be the prerequisite for many	is offered by one
Department	enrolls many	offers many	

From Table 6.2 we can see that a Student can take many courses and a Course has many students in the Student-Course relationship. A *cardinality ratio*, associated with each relationship, describes the multiplicity of the entities in a relationship. For example, the Student-Course relationship is many-to-many, or M:M, because one student can take many courses and one course is taken by many students. The relationship between Student and Department is M:1 since a Department enrolls many students, but a student can major in only one Department. A recursive relationship, prerequisite, exists between the Course entity and itself because a course may have many other courses as prerequisites and may be the prerequisite for many other courses. Thus the prerequisite relationship has an M:M cardinality. It needs to be noted that in this example the needs of the college required relationships between all pairs of entities. If, for example, the college did not need to keep track of the students' majors, then the relationship between Student and Department would be left blank in the entity relationship matrix.

The ERD constructed is shown in Figure 6.9. The relationships are identified by the diamonds shaped symbols; the entities are denoted by the rectangles with their name at the top. The cardinality ratio is labeled on the links between the relationships and entities. Another piece of information shown in the ERD is the attributes associated with each entity. An attribute is a feature or characteristic of an entity that needs to be remembered. There are many different types of attributes; however, the most important are *key attributes* (which are underlined) which uniquely identify instances. For example, the identification number (ID) attribute of a Student is a key attribute.

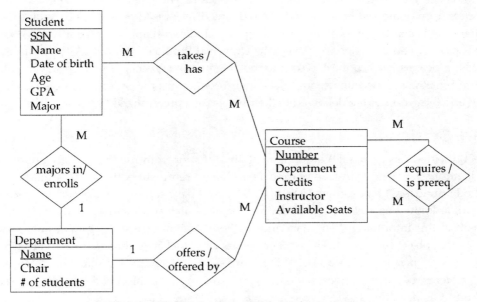

Figure 6.9 ERD for the college database system.

The ERD allows for an easy interpretation of the relationships between entities as well as their attributes. Although beyond the scope of the discussion here, the ERD is a formal language that can be used to automatically generate the database structure. The relationships in the ERD allow queries to be asked of the resulting database. For example, the college could derive a student's course schedule from the database from the relationship between Student and Course.

6.6 The Unified Modeling Language

The *Unified Modeling Language* (UML) was created to capture the best practices of the object-oriented software development process. However, it has valuable modeling tools that can be applied more generally to ECE systems. The objective of this section is to provide an overview of UML and the six different system views that it encompasses. A caveat for the reader—UML is complex, and complete coverage is beyond the scope of this book. In order to fully understand its subtleties and nuances, the reader is advised to consult the references provided in Section 6.8.

In order to aid in the explanation of the different views we will create a UML model of a system called the Virtual Grocer, or v-Grocer for short. The v-Grocer enables a user to order groceries from home and have them delivered. The concept is to provide users with a barcode scanner that is connected to their home computer along with application software. When the user runs out of an item, he/she scans in the Universal Product Code (UPC). When users want to order groceries, they connect to the Internet, log into the grocery store web server, enter the quantity for the scanned items, and place their order. Once the order is completed, they are billed and the groceries are delivered to their houses at a prearranged time.

6.6.1 Static View

Object-oriented design (OOD) is fundamentally different from functional software design in that it emphasizes objects instead of functions. *Objects* represent both data (attributes) and the methods (functions) that can act upon the data. An object represents a particular instance of something known as a *class*, which defines the attributes and methods. An object encapsulates all of this information. The data and methods that an object encapsulates are available only to that object by default. Thus, other objects cannot change that state of an object unless given specific permission to do so. This improves the reliability and maintainability of software systems, because changes made to the internal representation of a class are not seen by methods outside of the class.

The intention of the *static view* is to show the classes in a system and their relationships. The static view is characterized by a *class diagram*. A very simple class diagram with a single class is shown in Figure 6.10. The specification of a class has three parts, a name, a list of attributes, and a set of methods. The name of the class for this example is `Customer` and it has three attributes: `Name`, `Address`, and `CustId`. It has one method associated with it, `ChangeAddr()`, which can change the `Address` attribute. A class, just like an entity in an ERD, is a generalization of a set of a particular thing or instance. For example, the `Customer` class could have a particular instance called Ms. Robinson.

Customer
-Name : string
-Address : string
-CustId : long
+ChangeAddr() : bool

Figure 6.10 Class diagram notation. The class has a label (`Customer`), attributes (`Name`, `Address`, and `CustId`), and methods (`ChangeAddr()`).

Classes are related to one another by relationships that define how classes interact with each other. For example, if a class is a subset of another class (a kind of relation) then the subclass inherits all the attributes and methods of the *superclass*. In UML, there are a host of relationships; among these are generalization, composition, and associations. Two classes are related by a *generalization relationship* when one is a subset of the other. Two classes are related by a *composition relationship* when one is composed of members of the other. Two classes are related by an *association relationship* whenever they need to send messages to one another. Relationships are drawn as lines connecting the two participant classes. The terminals of the line have different shapes depending on the type of relationship. Just like an ERD, the relationships between classes have cardinality defined by the rule base. In order to illustrate these points, Figure 6.11 shows the class diagram for a portion of the v-Grocer system.

Figure 6.11 Class diagram for the v-Grocer.

The class diagram has five classes—`Customer`, `Delivery`, `Order`, `Item`, and `GroceryCart`. The `GroceryCart` class is derived from the many-to-many relationship which exists between the `Order` and `Item` classes. Just like the rules for an ERD, the cardinality of a relationship is read at the end of the association in which it is involved. The only difference is that the many cardinality in a class diagram is denoted by a * symbol. For example, a delivery is sent to one customer, while a customer may receive many deliveries.

6.6.2 Use-Case View

The intention of the *use-case view* is to capture the overall behavior of the system from the user's point of view and to describe cases in which the system will be used. It is characterized by a *use-case diagram*, and an example for the v-Grocer is shown in Figure 6.12. There are only two symbols employed in a use-case diagram, actors and use-cases. Actors, drawn as stick figure people, are idealized people or systems that interact with the system. For this example the actors are customers, delivery people, the database, clerks, and the web server. Use-cases are drawn as ovals in the diagram, and in this example, they are WebOrder, DeliverOrder, and AssembleOrder. A use-case is a particular situation when actors use the system and is usually represented by a sequence of actions that will be performed by the system. This sequence of actions typically represents a high level of functionality. Every actor that interacts with a particular use-case is connected to it with a line. Finally, the entire collection of use-cases is enclosed in a rectangle with the actors outside. This rectangle represents the system boundary and consequently is labeled with the name of the system.

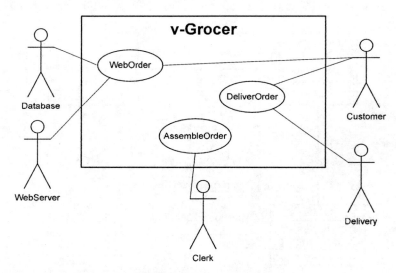

Figure 6.12 Use-case diagram for the v-Grocer.

From Figure 6.12 it is apparent that a Customer, the WebServer, and Database interact when creating a WebOrder. Use-cases are often described in a table as shown for this particular example in Table 6.3.

Table 6.3 `WebOrder` use-case description.

Use-Case	`WebOrder`
Actors	`Customer`, `Database`, and `WebServer`
Description	This use-case occurs when a customer submits an order via the `WebServer`. If it is a new customer, the `WebServer` prompts them to establish an account and their customer information is stored in the `Database` as a new entry. If they are an existing customer, they have the opportunity to update their personal information.
Stimulus	Customer order via the `GroceryCart`.
Response	Verify payment, availability of order items, and if successful trigger the `AssembleOrder` use-case.

Use-cases focus on a very high level of functionality and describe who will interact with the system and how they will interact. Given their high level of abstraction, they can easily be incorporated into a variety of ECE projects. They are also simple and easy to understand and thus can be incorporated into presentations to non-technical audiences. Finally, use-cases are simply not busy work—they are fundamental to the development of other UML models.

6.6.3 State Machine View

The *state machine view* is characterized by a *state diagram* as was discussed in Section 6.2. Again, the intention of a state diagram is to describe systems with memory. Figure 6.13 shows the state view of a customer logging into the v-Grocer web server.

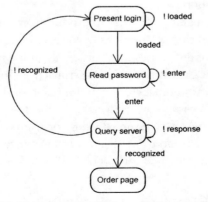

Figure 6.13 State diagram for the v-Grocer customer login.

6.6.4 Activity View

The intention of the *activity view*, characterized by an *activity diagram*, is to describe the sequencing of processes required to complete a task. In UML, the tasks elaborated are the individual use-cases identified in a use-case diagram. Since more than one actor may be involved in completing a task, an activity diagram can express the concurrent nature of tasks. An activity diagram is composed of states, transitions, forks, and joins. Figure 6.14 contains an activity diagram for the v-Grocer order delivery system. The diagram gives a clear visual picture for the activities that need to be completed for an order to be packed and delivered. After a complete order is placed, the flow is forked into two concurrent processing branches. One of these branches is for completion of the order, while the other addresses its scheduling, delivery, and coordination with other deliveries. When both of those processes are completed, they are joined together and then the delivery run is made.

Figure 6.14 Activity diagram for order processing and delivery for the v-Grocer.

6.6.5 Interaction View

The intention of the *interaction view* is to show the interaction between objects. It is characterized by collaboration and sequence diagrams. In an object-oriented system, tasks are completed by passing messages between objects. The interaction view shows how messages are exchanged in order to accomplish a task. These tasks are usually the use-cases. Since messages are sent through time, the interaction view must be able to express the concept of order. We start by examining collaboration diagrams.

Two objects collaborate together in order to produce some meaningful result. A *collaboration diagram* shows the sequencing of messages that are exchanged between classes in order to complete a task. It consists of the classes that participate in the realization of the task and the messages exchanged. The messages are drawn as arrows from the sending object to the receiving object. The message arrows are labeled with the name of the message and its numerical position in the sequence of messages exchanged to realize the task. For example, Figure 6.15 shows how the `WebOrder` use-case is realized using the `Customer`, `WebServer`, and `Database` classes. Note, `WebServer` and `Database` are introduced as new classes and are not shown as part of the class diagram in Figure 6.11.

Figure 6.15 Collaboration diagram for the WebOrder use-case.

The collaboration diagram is similar in form to the class diagram, the difference being that the relationships are annotated with the messages that are exchanged. Consequently, the collaboration diagram aides the developers in understanding and implementing the methods used between classes. In order to emphasize the order in which messages are exchanged, a developer can also use a sequence diagram.

A *sequence diagram* contains the same information as the corresponding collaboration diagram. Where the collaboration diagram emphasizes the objects that interact to produce a behavior, a sequence diagram emphasizes the message order that produces a behavior. As shown in Figure 6.16, the classes that participate in the behavior are listed in a row at the top of the diagram. From each class a dotted vertical line is drawn downward that represents the lifeline of its class (the vertical axis represents time). When an object is actively requesting or waiting for information from another object, a rectangle is drawn over the dotted lifeline of the object. The message is drawn as an arrow from the sending object to the receiving object and labeled with the name of the message. The activity diagrams can be applied generally to ECE systems to show the interaction between entities, particularly the sequencing of messages.

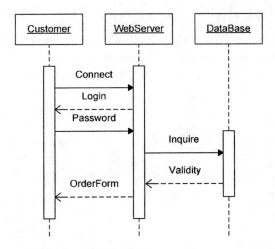

Figure 6.16 Sequence diagram for the WebOrder use-case.

6.6.6 Physical View

The intention of the *physical view* is to demonstrate the physical components of the system and how the logical views map to them. The physical view is characterized by a component and deployment diagram. A *component diagram* shows the software files and the interrelationships that make up the system. Software files are shown in rectangles. Lines connect together files that need to communicate. A *deployment diagram* shows the hardware and communications components that will be used to realize the system. The hardware components are drawn as cubes and labeled with their names. Figure 6.17 shows a combined component and deployment diagram for the v-Grocer system. The software files are the Browser, v-Grocer, Apache, and Oracle, while the hardware components are the Customer, WebServer, and Database.

Figure 6.17 The combined component and deployment diagram for the v-Grocer. The software files are the Browser, v-Grocer, Apache, and Oracle, while the hardware components are the Customer, WebServer, and Database.

6.7 Project Application: Selecting Models

Chapters 5 and 6 have presented a variety of models for representing the behavior of systems. The design team needs to select the correct combination of tools to properly describe a design for its eventual implementation. Table 6.4 provides a summary of the models and their different intentions.

Table 6.4 Guidance for model selection.

Model	Intention
Functional Decomposition	To describe the input, output, and functional transformations applied to information (electrical signals, bits, energy, etc.) in a system. It is broad in application. This often provides a view that is close to the actual system implementation. See Chapter 5.
State Diagram	To describe the behavior of systems with memory. They are very flexible when it comes to application. They are often applied to digital design, but state diagrams can also be used to describe the high level behavior of complex systems. The only prerequisite on their use is that the system has memory. See Section 6.2.
Flowchart	To describe a process or algorithm, including its steps and control. They are applicable to many problem domains from software to describing business practices. See Section 6.3.
Data Flow Diagram	To model the processing, transformation, and flow of data inside a system. They are typically supplemented by an event table describing all possible events and the resulting actions. Creating a DFD requires the designer to carefully think about the uses of the system and how the system is to react to external users and events. See Section 6.4.
Entity Relationship Diagram	To catalog a set of related objects (entities), their attributes, and the relationships between them. ERDs capture the entities and relationships of a portion of the world into a formal data model. The graphical language describes the attributes of the entities and the cardinality of the relationships. ERDs are formal enough to be unambiguously translated into a complete description of a database system. See Section 6.5.
	The Unified Modeling Language. The intention of UML is to describe complex software systems. However, certain views are well suited to describing systems at a high level and are applicable to many domains. The process of viewing a system from the six perspectives listed below decreases the chances that crucial details of the design will be overlooked.
Class Diagram	To describe classes and their relationship in an object-oriented software system. Class diagrams are primarily for software design and are not easily accessible to a non-technical audience. See Section 6.6.1.
Use-Case Diagram	To capture the overall behavior of the system from the user's point of view and describe cases in which the system will be used. See Section 6.6.2.
State Machine	This is essentially the same as the state diagram, but is also a formal UML view. See Sections 6.2 and 6.6.3.
Activity Diagram	To describe the sequencing of processes required to complete a task. Composed of states, transitions, forks, and joins. Can show concurrent processes. See Section 6.6.4.
Interaction Diagram	To show the interaction and passing of messages between entities in a system. Characterized by both collaboration and sequence diagrams. See Section 6.6.5.
Physical Diagram	To describe the arrangement and connections of the physical components that constitute a system. In the case of UML, it is characterized by a component and deployment diagram. See Section 6.6.6.

6.8 Summary and Further Reading

Models provide a convenient method of describing a system at a high level of abstraction. This allows a system to be described without having to determine all of the implementation details. All models are not the same—they come in a variety of forms and each is designed to serve a different intention. The properties of effective models were presented, and we saw that they are similar to those of an engineering requirement. Models should encourage innovation by allowing the exploration of alternative implementations. Since models are built from abstractions of the actual system components, they are an effective means for communicating with non-technical stakeholders. Finally, models provide an excellent means of documenting the development of a design from the highest level down to the detailed design.

This chapter presented a variety of models for describing system behavior that included state diagrams, flowcharts, data flow diagrams, entity relationship diagrams, and the Unified Modeling Language. Table 6.4 provides a quick reference that describes the intention and application of the models examined in both Chapters 5 and 6.

Flowcharts have been around for quite some time and an early work describing them is Flowcharts by Chapin [Cha71]. The book Programming Logic and Design [Far02] covers flowcharts extensively and their application in programming. State diagrams are fundamental to the ECE field and further information on them is available in virtually any introductory digital design textbook. Data flow diagrams and entity relationship diagrams are common tools used in information systems analysis and design and can be further explored in Systems Analysis and Design [Sat02] and Fundamentals of Database Systems [Elm94]. Two references for UML are The Unified Modeling Language Reference Manual [Rum98] and Schaum's Outline of UML [Ben01].

6.9 Problems

6.1   Why is it important for a model to separate the design of a system from its realization?

6.2   Classify each of the following as either a model, not a model, or sometimes a model. Justify your answer.

a)   A diagram of a subway system.
b)   A computer program.
c)   A football play.
d)   A floor plan of the local shopping mall (a "you are here" diagram).
e)   An equation.

6.3   Which of the following systems has memory? Justify your answer using the concepts of input, output, and state.

a)   An ink pen.

b)   A resistor.

c)   A capacitor.

d)   A push button.

6.4   A can of soda has memory. Your objective is to figure out what characteristic of the can is the state variable and what input causes it to change. Based upon this, draw a state diagram for a can of soda. Label the transition arcs with the input responsible for the transition. Hint: no special equipment is needed to elicit the change of state.

6.5   Consider the state diagram for the vending machine shown in Figure 6.2. Now assume that the system accepts nickels, dimes, and quarters. Also assume that it is capable of returning change to the user after a purchase. Create a state diagram that represents this new system. Make sure to define the output signals and their value for each state.

6.6   Use a state diagram to describe the high level operation of the ChipMunk Recorder (CMR). The CMR records sounds and then plays them back at a variety of speeds, making a recorded voice sound like a high pitched chipmunk. The CMR receives user input from a keyboard and an audio source. The behavior of the system is described as follows:

- When powered up, the CMR enters a wait state.

- If 'R' is pressed the recorder begins recording.

- Any key-press will put the CMR back into the wait state.

- If 'S' is pressed, the CMR is ready to change the playback speed. A subsequent numerical input between 1 and 5 will cause the playback speed to be changed to that value.

- Pressing the 'R' key when in the adjust playback speed mode will cause the CMR to go to the wait state.

- Pressing a 'P' key will cause the CMR to playback the recorded sounds. When done playing the entire recording, the CMR will loop back and start playing at the beginning.

- Any key-press while in the playback mode will cause the CMR to go back into the wait state.

Draw a state diagram describing the behavior of the CMR. Write a brief description of the output inside each state.

6.7   Build a flowchart to describe the operation of a microcontroller (MCU) based temperature regulating system. The system monitors the temperature of a heated environment using thermistors and regulates the temperature by turning on fans. The MCU periodically reads the temperature from 64 different thermistors (each is driven by its own constant current source) by selecting each through an analog multiplexor. The resulting voltage is converted into an 8-bit digital value using the MCU's analog-to-digital converter. There are two temperature ranges recognized—low temperature and working temperature. When the temperature is low, the MCU turns all the fans off. When the temperature is in the working range, the MCU uses a complex algorithm to determine the number of fans to turn on.

6.8   Build a data flow diagram and event table to represent an image archiving system for an art museum. The art museum maintains a database of digital images of paintings from museums all over the world. The following information is known about the image database system:

- Images are shared among museums in a participating network. Whenever a participating museum posts a new image, it sends a broadcast email to all participating museums in the network with the image attached as an email. It provides the name of the painting and artist in the body of the email. All new images received are added directly to the museum's own image database.

- When inserted into the museum's database, each image is provided with a tag identifying the name of the painting and the artist. Furthermore, this triggers an image analysis routine that classifies the image based into predefined categories such as portrait, natural scene, and modern. Furthermore, it stores key features that are extracted from the image.

- The key features are used to identify and retrieve visually similar images from the museum's database. Another image processing algorithm is run that compares the visual similarity of the new image to all images in the database. This process produces a matching score of 0 to 100 that is stored.

- The museum's image database is available to visitors via computer kiosks placed throughout the museum. Kiosk users can retrieve and view images in one of three ways. First, they can specify the name of the artist or painting. Second, they can retrieve a class of images, such as modern. Third, once they have received a painting, they can submit a request to view visually similar images. The visually similar images are retrieved for viewing based upon the matching score.

6.9   Build an ERD to keep track of the bicycle frames manufactured at a local company. The following are notes from an interview with the owner:

- When a customer comes in to place an order they are measured for a frame size. The measurements made are leg, torso, and arm length.

- A customer's name, address, and phone number must be stored.

- The customer must also choose one of three types of frames: X, Conventional, and Suspension.

- The customer can also choose a frame color.

- The frame's unique serial number must be stored.
- The tubing attributes used to construct each frame must be stored. Tubing has the following attributes: type, lot #, OD, ID, and manufacturer.

6.10 **Project Application.** Develop behavior models that are applicable for describing your system. Table 6.4 is provided to help in making the determination as to which models are applicable.

Chapter 7 Testing

A stitch in time saves nine.—Anonymous

Most systems undergo testing throughout their development and before they are delivered to the customer. Clearly, systems should be tested to ensure that they meet the requirements developed in Chapter 3. In fact, one of the desirable properties of an engineering requirement is that it be verifiable, in other words, testable. The philosophy of testing is embodied in the quote above, which means that it is better to correct errors early, rather than wait until they become much larger problems later. As we saw in Chapter 1, the cost to correct problems increases exponentially with the lifetime of the project. Thus, testing should be considered throughout system development.

This chapter presents fundamental testing concepts and describes how they are applied throughout the development lifecycle. Concepts of controllability and observability are addressed since they impact the ability to conduct and interpret the results of a test. Approaches to debugging systems are provided, followed by unit-level testing, integration testing, and acceptance testing. This chapter also provides templates for planning and documenting tests so that they may be applied to design projects.

Learning Objectives

By the end of this chapter, the reader should:

- Understand the concepts of black box tests, white box tests, observability, and controllability.
- Understand debugging processes and the difference between Heisenbugs and Bohrbugs.
- Understand when a unit test is used and how it is constructed.
- Understand when an integration test is used and how it is constructed.
- Understand when an acceptance test is used and how it is constructed.

7.1 Testing Principles

The design process increases in specificity from development of the engineering requirements to the detailed design. Testing begins once construction of the system starts. The test process works in the reverse direction, from the most detailed level of the design to the most abstract. This process is shown in Figure 7.1, which splits the design process into two equal halves, development and testing. Testing typically starts with low-level debugging of the components as they are developed, followed by unit testing of the components, integration testing of subsystems, and finally, acceptance testing of the entire system. Each of these tests phases has a corresponding phase on the development side in which the testing benchmarks are set. It is common for the testing process to occur in parallel with other phases of development. For example, acceptance test plans are typically created in conjunction with the Requirements Specification.

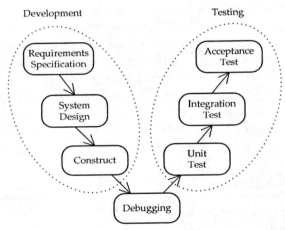

Figure 7.1 The Test Vee. Design stages are on the left and corresponding tests are on the right.

Testing is a way to check that the actual behavior of a component or system matches the expected behavior. It seems obvious that, in order to test a system, you must have a thorough knowledge of the system's expected behavior. All too often, engineers perform system testing by turning on a system to see if it works. Without knowledge of the expected behavior and what is supposed to happen inside the system, it is impossible to ascertain the operational state. Consequently, it is important to have a deep understanding of the expected operation of the system and its components before testing begins. A way to achieve this understanding is to develop a clear design document, as advocated in Chapters 5 and 6, with detailed information about the intended behavior of system components. The writing process helps to organize the designer's thoughts about the system, document design decisions for others to understand, and force the designer to look at the system from a different perspective.

How should testing be done? Given enough time, a system could be tested by simply submitting random inputs and observing the outputs. This process is not efficient because it would take an inordinate amount of time to uncover errors. Tests should be chosen so that they maximize the likelihood of finding errors. Ideally, tests should be developed in parallel with the design of the components to ensure that the test cases developed provide adequate coverage.

7.1.1 Types of Testing, Observability, and Controllability

Tests fall into two general types—black box and white box tests. *Black box tests* are those that are performed without any internal knowledge of the system. In a black box test, the testing is typically conducted by changing the inputs and observing the system outputs. *White box tests* are conducted with knowledge of the internal working of the system. It is possible that internal nodes of the system may be changed and observed. However, as we will see in later examples, knowledge of internal operation may simply influence how the test is constructed.

One of the many goals in designing a system is to increase its testability. A design is *testable* when a failure of a component or subsystem can be quickly located. A testable design is easier to debug, manufacture, and service in the field. One way to increase the testability of a system is to increase controllability and observability. *Controllability* is the ability to set any node of the system to a prescribed value. *Observability* is the ability to observe any node of a system. In black box testing, both controllability and observability are low. In white box testing, controllability and observability may be higher depending on the design.

Let's examine this further via the example of a simple transistor amplifier shown in Figure 7.2. The purpose of this circuit, known as the common-emitter amplifier, is to amplify the input signal, v_i, to produce a linearly proportional output signal, $v_0 = A \times v_i$. The rectangular boundary in the figure represents a black box view of the system. In this view, the system power, V_{cc}, and ground would be applied to activate the circuit. The black box testing would consist of checking supply and ground voltages, varying the input signal, and observing the output signal. Again, this is a low controllability and low observability situation.

White box testing utilizes knowledge of the internal workings of the design. When designing a transistor amplifier, there are two major points to consider—the DC bias voltages in the circuit and its AC, or time varying, amplification behavior. The two behaviors are related since the AC behavior depends upon proper DC biasing of the circuit. During detailed circuit design, the expected DC voltages for different nodes in the circuit would be determined. Thus, a white box test would consist of first checking the power supply and ground voltages as was done in the black box case. The next step would be different in that the node voltages (V_B, V_C, V_E) would be checked to see if they meet the expected design values. This indicates a high degree of observability. However, the controllability is not significantly better than in the black box case. This is because the internal DC node voltages in the circuit cannot arbitrarily be changed without negatively changing the operation of the circuit.

Figure 7.2 Transistor amplifier design.

Another demonstration of controllability and observability is that of finite state machines in digital systems. *Finite state machines* are the hardware realization of the state diagrams that were examined in Chapter 6. Figure 7.3 (a) shows the general structure of a finite state machine. It has a combinational logic block (realized using AND, OR, etc. gates) and a feedback section containing sequential logic, which in this particular instance are D flip-flops. The feedback loop provides the ability for the system to have memory that is characteristic of state machines. Let's assume a white box testing scenario, where there is access to all internal nodes of the system. The assumption would be that the system has a high degree of observability, but it may come as a surprise to find that is not true! This is because system behavior is defined by its state diagram—the design may limit the number of states that the system may enter. Thus, if you want to observe node values for an arbitrary state, you are only able to do so if it is designed into the system. Furthermore, since the system has memory, you need be able to control prior states to observe a particular behavior that you want to see. Therefore, it also has low controllability.

Both the controllability and observability of the finite state machine can be improved by the inclusion of additional hardware as shown in Figure 7.3 (b). In this case, two multiplexors (MUXes) are added in the feedback loop. The select pin on the MUXes allows selection of either inputs Y0 and Y1 which are then fed to the D flip-flops. This allows the system to be put into an arbitrary state for testing, which gives a high degree of controllability and obervability.

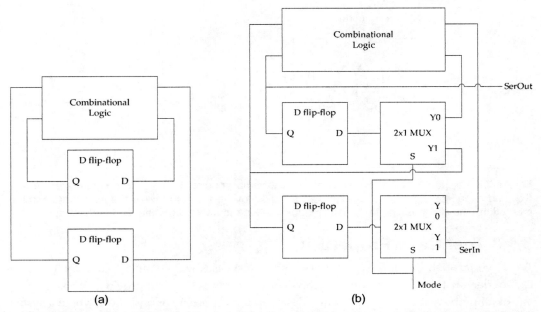

Figure 7.3 (a) The general structure of a finite state machine with two D flip-flops in the feedback loop. (b) The finite state machine containing multiplexors. This provides increased controllability and observability.

7.1.2 Stubs

Stubs are valuable aids for testing, and are commonly applied in software testing. A *stub* is a device that is used to simulate a subcomponent of a system. Typically, stubs are used to simulate inputs or monitor outputs of a circuit being tested, the *unit under test*. Stubs are used to ensure that a subsystem operates correctly before connecting it together with other parts of the system. This saves debugging time and may prevent damage to other subsystems. Inclusion of stubs should be considered when in the detailed design phase. In software design, stub routines are developed to either call other functions or act as those to be called by the unit under test.

Consider a hardware example, the transistor amplifier in Figure 7.2. Assume that the circuit is ultimately to be integrated into a larger system. The input to this system is to be a time-varying source with certain resistive and inductive characteristics, while the output will be connected to another system with a known input resistance range. The stubs used for testing in this system are shown in Figure 7.4. On the input side is a function generator, an off-the-shelf component, connected to a resistor and inductor that model the expected characteristics of the final system. The stub on the output side is simply a resistor, whose value can be varied over the expected output range.

Figure 7.4 The use of stubs for testing a transistor amplifier circuit. The function generator, resistor (R), and capacitor (C) model the expected behavior of the input source in the final system implementation. The variable resistance (R_L) models the load that would be attached to the output.

7.1.3 Test Case Properties

As we go through the different levels of testing, test cases will be built. Generally, test cases cannot demonstrate that a system operates correctly unless they can provide complete coverage of all problems. Test cases are often designed to find problems and show where a system is not operating properly. The properties of an effective test case are that it should be [Run99]:

- *Accurate.* The test should check what it is supposed to and exercise an area of intent.
- *Economical.* The test should be performed in a minimal number of steps.
- *Limited in complexity.* Tests should consist of a moderate number (10-15) of steps.
- *Repeatable.* The test should be able to be performed and repeated by another person.
- *Appropriate.* The complexity of the test should be such that it is able to be performed by other individuals who are assigned the testing task.
- *Traceable.* The test should verify a specific requirement. The corresponding requirements for the different types of test are derived from the associated development stages in the Test Vee in Figure 7.1.
- *Self cleaning.* The system should return to the pre-test state after the test is complete.

7.2 Debugging

At some point in the design process, the implementation level must be reached, where tasks such as constructing circuits, wiring integrated circuits, and writing code are carried out. Applying the functional decomposition paradigm introduced in Chapter 5 should provide a clear idea of the inputs, outputs, and behavior of the modules that are being built. Inevitably, there will come a point during the construction of a component when it will not function as expected. This is commonly referred to as a *bug*. It requires the application of debugging skills to determine the root cause of the problem and correct it. You have undoubtedly run across a variety of bugs in your day, and it is a good guess that your bugs fell into one of two camps—Bohrbugs and Heisenbugs.

Bohrbugs are named after the Bohr model of the atom that assumes that electrons have a distinct position in space. Bohrbugs are reliable bugs, in which the error is always in the same place. This is analogous to the electrons having a definite position. Given a particular input, a Bohrbug will always manifest itself in the same way and in the same place. Finding a Bohrbug is a matter of laying the correct trap. A good trap is simple to set-up, quickly causes an error, and reveals the source of the error. This is a tall order, but one which experience hones.

Heisenbugs are named after the Heisenberg Uncertainty Principle, in which the position of an electron is uncertain. Analogously, Heisenbugs may not always be reproducible with the same input. They seemingly move around within a system and are consequently difficult to locate. Finding a Heisenbug requires you to think outside the box because they usually result from unanticipated mechanisms. An example of a Heisenbug is a computer program with a pointer error that occasionally overwrites the system stack. This can cause return values from a subroutine to be incorrect. In such a case, the subroutine would appear to have a problem, since it is returning the wrong value. However, testing the subroutine by itself would confirm that the subroutine works properly. Another good example is a circuit that works fine on some days, but doesn't work on others (typically when a professor is nearby). Insidious problems such as a floating ground line often are to blame.

Regardless of the bug type, the debugging process is iterative. You must run tests and depending on the results, go back and run new tests. With this in mind, you should enter into the debugging process with a strategy in mind. This strategy is often similar to programming an if-then structure—"if the test is negative, then I'll pursue this line of attack; otherwise the error could be in another subsystem." In general, the debugging process is much the same as the process of scientific understanding. The steps of the debugging process are:

- *Observe.* Observe the problem under different operating conditions.
- *Hypothesize.* Form a hypothesis as to what the potential problem is.
- *Experiment.* Conduct experiments to confirm or eliminate the hypothesized source of the problem.
- *Repeat.* Repeat until the problem is eliminated.

When hypothesizing, make sure to check the simplest and easiest problems first. There are two good reasons for this—they are easy to perform and more tests can be performed in a given period of time. In addition, designs should be verified from the lowest levels of abstraction to the highest. For example, voltages should be verified as correct before moving to higher levels of functionality. The reason for this heuristic is obvious—the higher level of functionality cannot operate correctly unless all the lower levels are working.

7.3 Unit Testing

A *unit test* is a test of the functionality of a system module in isolation, and the unit test should be traceable to the detailed design. A unit test consists of a set of test cases used to establish that a subsystem performs a single unit of functionality to some specification. Test cases should be written with the express intent of uncovering undiscovered defects.

In order to facilitate unit testing it is best to design units that are highly cohesive and loosely coupled (see Chapter 5, Section 5.8). High cohesion means that the unit does a few things very well; loose coupling means that it has few connections. Units with both attributes are testable because they can easily be separated from the system and given targeted inputs.

The best time to write tests is while the unit is being designed. There are two significant advantages to this approach of test writing. First, the process of thinking about test situations and conditions may lead the designer to uncover errors before they are ever designed into the system. Second, white box unit tests need to be written with an understanding of the internal organization of the component, and a system is best understood during its development.

For example, consider the design of a piece of hardware to convert temperatures in Celsius to Fahrenheit. Let the operation of a digital hardware design be represented by the following pseudo-code.

```
if (16 < Celsius Temperature < 32)
    Fahrenheit Temperature = ROM[input-16];
else
    Fahrenheit Temperature = (9* Celsius Temperature)/5 + 32;
```

When the input temperature is between 16 and 32 degrees, the output temperature is determined from a lookup operation in the ROM (read-only memory); otherwise, it is converted using the familiar Celsius to Fahrenheit conversion (yes, this is an inefficient way to convert temperatures). Knowing this algorithm would certainly affect the choice of tests. For example, input temperatures in the range [17, 31] would be selected in order to test operation of the ROM, while values outside this range would be selected to test the arithmetic operation. Additionally, a set of input temperatures {16, 17, 31, 32} could be selected to test the inequalities that determine the source of the conversion.

This example illustrates the concept of a *processing path*, which is a sequence of consecutive instructions or states encountered while performing a computation. The temperature conversion example has two processing paths, one where the *if* statement is taken and one when the *else* statement is taken.

Each path through the system represents a potential test case. The extent to which the test cases cover all possible processing paths is called the *test coverage*. It is desirable to design test sets that have the highest coverage as possible. The ultimate in coverage is achieved by *path-complete coverage* where every possible path has a test. However this level of coverage may not be possible. One reason is that the number of paths goes up exponentially with the number of branches (i.e., *if* statements), since each additional branch doubles the number of processing paths. In cases where there are more paths than it is possible to check, you must be satisfied with partial path coverage. In such cases, those paths that are most likely to reveal an error in the design should be tested.

The resulting collection of paths forms the unit test. The unit tests and their accompanying documentation contain important information about the behavior and organization of a component. In addition to serving as a documentation tool, the unit test serves as a communication tool. Unit tests are generally written by one person or group, while another person or group performs the tests. This is done to eliminate the possibility of bias during the testing process. In order to help organize the process of running the unit tests and preserve the unit tests for later use, the individual test cases must be organized. Test cases can be organized using matrices, automated scripts, or step-by-step documents.

7.3.1 Matrix Tests

A *matrix test* is best suited to cases where the inputs submitted are structurally the same and differ only in their values. For example, consider a test for the analog-to-digital converter (ADC) that was used in the temperature measuring system in Chapter 5 (Section 5.7, Figure 5.11). From the accompanying functional description of the design, we know that the output has 10-bits and that the converter needs a clock. Assume that the clock frequency is 10kHz and that the ADC responds to erroneous clock signals by asserting zero on its output. Further, assume that the input voltage range varies from zero to five volts.

Since an ADC is an integrated circuit, a black box test is appropriate, meaning that the test values must be constructed without knowledge of the internal organization. The unit test consists of submitting different inputs to the ADC and verifying the outputs. Since these tests will only present variations on the input, with no change in the testing procedure, a test matrix is created. A test matrix template for this example is shown in Table 7.1.

Blank items in the matrix are filled out by the person performing the test. Hence, the test writer must take care in organizing the information on the form because someone else will use it to run the test. The test presented in Table 7.1 tests two requirements—does the ADC convert voltages correctly with a functioning clock and does it operate correctly with a non-functioning clock?

Table 7.1 A matrix test template for an analog-to-digital converter.

Test Writer: Sue L. Engineer								
Test Case Name:	ADC function test				**Test ID #:**		ADC-FT-01	
Description:	Verify conversion range and clock frequency. Output goes to 0 in presence of null clock.				**Type:**		☐ white box ☑ black box	
Tester Information								
Name of Tester:					**Date:**			
Hardware Ver:	1.0				**Time:**			
Setup:	Isolate the ADC from the system by removing configuration jumpers.							
Test	V_T	Clock	Expected output		Pass	Fail	N/A	Comments
			Decimal	Hexadecimal				
1	0.0V	10kHz	0	0x000				
2	2.0V	10kHz	410	0x19A				
3	5.0V	10kHz	1023	0x3FF				
4	5.0V	0Hz	0	0x000				
5	2.0V	0Hz	0	0x000				
Overall test result:								

7.3.2 Automated Script Tests

An *automated script test* is a sequence of commands given to a unit under test. For example, a test may consist of a sequence of inputs that are provided to the unit. The outputs for each input are then verified against pre-specified values. An example of an automated script is the memory self-check that occurs when a computer starts. In such a test, specific data is written sequentially to memory locations and the data is then read back from the locations to verify whether it matches the value written. Automated scripts are also common in hardware simulation software packages that provide the capability to submit a range of inputs for testing. The specific inputs used by an automated script can be determined using path coverage techniques or to examine specific boundary conditions.

7.3.3 Step-by-Step Tests

A *step-by-step test* case is a prescription for generating the test and checking the results. These descriptions are most effective when the test consists of a complex sequence of steps. The test template for a step-by-step test has the all information contained in the matrix test template. As an example, recall the state diagram for the vending machine in Chapter 6 (Figure 6.2) that accepts nickels and dimes and dispenses candy when a total of $0.25 (or more) is submitted.

The state machine has different processing paths depending upon the combination and order of coins deposited. Test cases can be written for each of the processing paths through the system.

An example step-by-step test is shown in Table 7.2 for one particular processing path. The test checks the requirement that the vending machine dispenses candy when a total of $0.30 is deposited (in the design in Chapter 6, ≥ $0.25 is required to dispense and no change is returned). This is a white box test because the internal operation of the system was used to determine the test pattern. Additional tests would be created to test other combinations of inputs.

Table 7.2 A step-by-step unit test for the state machine vending application.

Test Writer: Sue L. Engineer							
Test Case Name:	Finite State Machine Path Test #1		Test ID #:		FSM-Path-01		
Description:	Simulate insertion of money with a mix of nickels and dimes. Verifies FSM, outputs candy in response to a total deposit of $0.30.		Type:		☑ white box ☐ black box		
Tester Information							
Name of Tester:			Date:				
Hardware Ver:	1.0		Time:				
Setup:	Make sure that the system was reset sometime prior and is in state $0.00.						
Step	Action	Expected Result	Pass	Fail	N/A	Comments	
1	Strobe Nickel	State should go to $0.05					
2	Strobe Dime	State should go to $0.15					
3	Wait	State should remain $0.15					
4	Strobe Nickel	State should go to $0.20					
5	Strobe Dime	State should go to $0.25					
6	Nothing	State should go to $0.00					
	Overall test result:						

7.4 Integration Testing

After the units of a system have been constructed and tested, they must then be integrated into larger subcomponents leading eventually to the construction of the entire system. The intermediate subsystems must be tested to make sure that components operate correctly together. The purpose of *integration testing* is to identify errors in the interactions of subsystems.

The test cases for integration testing are traceable to the high-level design, and the test cases are written based on characteristics of the design architecture. Potential test cases for integration can be derived from the following questions:

- What are the different paths of execution through the system?
- Are all modules exercised at least once during integration testing?
- Have all the interface signals been tested?
- Have all interface modes been exercised?
- Does the system process information at the required rate and meet timing requirements?

7.5 Acceptance Testing

An *acceptance test* is a formal document stipulating the conditions under which the customer will accept the system. An acceptance test verifies that the system meets the Requirements Specification that was developed in Chapter 3. Ideally, the acceptance test plan is developed with the engineering requirements and is traceable to them. It should identify the scope of the tests, that is, how much will be tested, the testing approach, and to what level the acceptance test will be taken to (unit or integration).

7.6 Application: Robot Design

In order to demonstrate the concept of testing for an application, consider the design of a robot that navigates autonomously. The objective of the robot is to move down a hallway. The engineering requirements for the robot are the following:

- *It has to move parallel to a wall while staying within 12 to 18 inches of it.*
- *Its heading should deviate no more than 10 degrees from the wall's axis.*
- *It should be able to operate under a variety of ambient lighting conditions.*

To meet these requirements, assume that the high level design in Figure 7.5 was developed.

Figure 7.5 Level 1 design architecture for the mobile robot.

The heart of the design is a microcontroller (MCU). The MCU reads the sensor values, makes decisions, and controls the outputs. The outputs of the MCU go to a differential drive train that consists of two independently powered DC motors that drive the wheels. The robot moves and turns by adjusting the relative speed of each motor. The MCU controls each drive motor through its own H-bridge using a pulse-width modulated (PWM) signal. H-bridges are circuits that transfer power to the motors and allow the direction of rotation to be changed. The duty cycle of the PWM signal controls the speed of the motor—a 0% duty cycle signal stops the motor and a 100% duty cycle signal runs the motor at full speed. The MCU also outputs a set of signals to control an informational display LCD, increasing the observability of the entire system. The signal to the digital compass is bidirectional because the MCU must configure the operating mode of the compass before using it. The MCU receives an analog input from the range finder where the voltage level of the signal is proportional to the distance to an obstacle in front of the sensor. Finally, a set of switches are included that allow quick re-configuration of the MCU for testing purposes.

The transformation from sensor inputs to motor outputs is controlled by the algorithm inside the MCU. This algorithm reads the range finder and compass, checks their values, and if they are within the pre-specified limits, the MCU does not change the state of the motors. If readings exceed the limits, the MCU breaks out of the monitoring loop and takes corrective action. This is done by adjusting the power to the motors until the sensor values are within their pre-specified limits.

7.6.1 Unit Testing

The design architecture of the system in Figure 7.5 contains seven units (digital compass, range finder, switches, MCU, LCD, H-bridges, and DC motors) to consider for unit testing. Let's consider each of the components in turn and examine how to construct unit tests for each under the three engineering requirements for hallway navigation. The MCU is used as part of the unit test for each of the other modules. This is because in itself it is a useful platform that can be used to perform the unit tests. Strictly speaking, this would not qualify as completely independent tests of each unit. However we believe that it's better to be pragmatic about testing rather than dogmatic. That is, it is better to create a flexible testing platform using a combination of the subsystems, such as the MCU and LCD, rather than to test these pieces in isolation. Each of the unit tests descriptions given below is general and should be formalized using the templates Tables 7.1 and 7.2.

MCU: As a first step, a unit test should be conducted on the MCU to ensure that it operates. This would likely be in the form of the execution of a simple program, e.g., "Hello World."

LCD: The LCD should have a unit test to exercise all internal LCD memory locations and display them. If the LCD is required to change modes (scrolling, for example), then each mode should be given its own unit test.

Switches: A unit test should be conducted to make sure the switches operate using a multimeter. It would also make sense to write a simple MCU program to display the switch settings on the LCD.

Digital Compass: A unit test to be conducted with the compass would be to demonstrate the establishment of communication between it and the MCU. This would require the transmission of configuration data to the compass and receiving of data from it. A second unit test would be to verify the functionality of the compass. This would require a method of changing the direction of the robot to a known orientation, and comparing this to the compass reading. The desired accuracy of the system would have to be taken into consideration.

It is important to note that the effort expended in developing this unit test is not wasted — in fact there are two very useful applications for this test. The first is that the subroutines written for this unit test should be reusable in the final design! The second application comes when the robot starts operating incorrectly, as all robots have a tendency to do. When this occurs, the digital compass unit test can be used to check the operation of the compass to eliminate or identify it as the source of the problem.

Range Finder: A unit test of the range finder consists of recording the actual distance between the sensor and a wall versus the distance measured by the output voltage. Since the range finder does not need any configuration information, the unit test could be run independently of the MCU. Additional tests should be run to determine the effects of any anticipated environmental variations that the robot may encounter, since infrared range finders are sensitive to variations in ambient lighting.

H-bridge: The H-bridge should be unit tested under a variety of loads and under the full range of operating modes (forward, reverse and stop).

Motor: The motor must produce an adequate torque to move the robot, and the unit test should verify that the motors produce enough torque to maneuver the robot chassis.

MCU (again): A processing path analysis of the control and navigation algorithm should be performed.

7.6.2 Integration Testing

There are many interactions between subsystems that could be tested during integration testing. A careful examination of the system must be done to determine which combinations of subsystems are most likely to reveal problems. Experience and component selection play a large part in molding expectations. For example, some infrared range finders have the bad habit of inducing large amounts of noise into the system. This knowledge would result in an integration test to check that this noise does not adversely affect the MCU. Each of the integration test descriptions given below is general and should be formalized using the templates in Tables 7.1 and 7.2.

MCU + DC motors + H-bridge + Switches: The DC motors can have an indirect effect on all the other subsystems through induced electrical noise generated during the switching of large currents using a PWM schema. The integration test should check the movement of the robot over the range of expected speeds. There should also be tests at the boundaries of full motor stop and full motor power under load. The switches can be used to change the conditions.

Chassis + Digital Compass + MCU + DC motors + H-bridge + LCD: The integration of the compass to the MCU was addressed in the unit test. However, the chassis and the electro-magnetic fields generated by the DC motors can adversely impact the compass operation. A test would consist of rotating the robot to known orientations and querying the compass at each orientation. Several tests could be performed under different motor speeds, loads, and directions. These motor parameters could be configured using the switches.

Range finder + Chassis + MCU + DC-motor + H-bridge: The effect of the noise generated by the DC motors on the infrared range finder should be examined. The integration test should consist of the range finder mounted on the mobile robot chassis placed at various distance from a wall. The test itself would consist of plotting the actual distance to the wall against the distance measured by the MCU. Several tests could be performed under different motor speeds, loads, and directions. Again, the motor parameters could easily be configured using the switches.

7.6.3 Acceptance Testing

The acceptance test is aided by the fact that the three engineering requirements set forth are verifiable. Several tests could be performed to verify that the robot responds correctly to environmental conditions. In the first case, the robot chassis could be mounted on a platform that could be rotated, allowing the drive wheels to move freely off the ground. The robot could be rotated and its responses recorded and compared against those actions necessary for corrective actions. The LCD could be used to aid the test by displaying the current state of the drive wheels and sensor values. A second test would involve testing its ability to stay within the specified distance from a wall. It could be moved to different positions from a wall and the responses of the drive wheels recorded and compared against those actions necessary for corrective reactions. Another test could be performed to measure the effect of ambient light variations.

The final set of acceptance tests would consist of observing the robot in an actual navigation situation. Lines could be placed on the floor in order to check that the robot maintains the correct distance from the wall. Several tests should explicitly perturb the robot to verify that it can take corrective actions.

7.7 Guidance

Tests have a lifetime beyond the obvious need to check proper operation of the subsystems, their integration, and the overall performance of the system. Test cases describe how the system operates in plain English. Test cases can be used to develop diagnostics, assist in writing technical documentation, and aid marketing and sales staff in understanding system performance. Testing is a value-added process in design.

Beyond attempting to remove bugs from the system, Burke and Coyner [Bur03] suggest the following are good reasons to perform testing:

- *Testing reduces the number of bugs in existing and new features.* Testing does not eliminate all the bugs, but rather reduces the probability of a bug making it to production.

- *Tests are good documentation.* Tests provide insight to others on the operation of the unit under test and how to interface to it.

- *Tests reduce the costs of change.* A change to a complex design with no tests can produce bugs that are difficult to track down. A good test plan can help localize the effect of bugs introduced by changes.

- *Tests improve design.* In order to create a testable design, you need to create highly cohesive, loosely coupled units.

- *Tests allow you to refactor.* Subcomponents of a testable design can be changed and optimized with less chance of introducing new errors. This is because tests exist that can verify the redesigned (refactored) module functions correctly.

- *Tests constrain features.* When a test is written before building the associated module, the exact requirements are defined. Hence, when a unit passes its test, there is confidence that the requirements have been met.

- *Tests defend against other designers.* Often a design needs to have circuitry to deal with special cases. Tests that check these special cases can make sure that future modification does not remove them.

- *Testing is fun.* Writing tests requires creative solutions to complex design problems.

- *Testing forces you to slow down and think.* When writing a test before incorporating a feature into a design, you are forced to see how the new feature fits into the existing design framework.

- *Testing makes development faster.* On a component level, testing slows development. However, as the design becomes larger and more complex, modules can be more easily integrated into the design without causing malfunctions in existing components.

- *Tests reduce fear.* Would you rather improve a unit with a test suite or one without?

7.8 Summary and Further Reading

Testing is an important part of the design process that helps to ensure systems will operate properly. This chapter examined basic principles of testing including black box testing, white box testing, controllability, and observability. They address the manner in which tests can be conducted, controlled, and states of the system observed. The use of stubs, which are employed to simulate system inputs and outputs were examined, as well as the properties of test cases. The different phases of testing from unit tests to integration tests to acceptance tests were examined. Testing proceeds from the most detailed level of the system to the most general, and the tests performed in each phase are traceable to their corresponding phases in the design development process.

The field of testing has been well developed by the software engineering community. Software Engineering: An Engineering Approach [Pet00] provides a good overview of testing principles such as black box and white box testing. It also includes a number of test strategies beyond those considered here. The Glossary of Vulnerability Testing Terminology from the University of Oulu's Electrical and Information Engineering department [Oul04] provides an extensive list of terms related to testing in the software domain. Gray's 1985 article *Why Do Computers Stop and What Can Be Done About It?* [Gra85] coined the terms Heisenbug and Bohrbug. This article introduces many interesting facts about how supercomputers fail. It provides a rare chance to look at the inner world of a supercomputer company. Many of the topics in the unit test section were influenced by Dianne L. Runnel's article *How to Write Better Test Cases* [Run99]. In this article, she defines precisely what is meant by unit test, and gives a clear picture of how to construct a unit test. An exceptional set of documents and templates are available from the Systems Engineering Processing Group of the United States Air Force. While intended for software development, the checklists for unit and integration testing contain many insightful points. They are accessible at
https://web1.ssg.gunter.af.mil/sep/SEP/MENUS/PD_LX009.asp .

7.9 Problems

7.1 Explain the differences between black box and white box testing.

7.2 Identify a circuit simulator (analog or digital) that you are familiar with. Explain the features of this simulator, which increase the observability and controllability of the circuit being simulated.

7.3 A mobile robot is being built. It uses a two DC motors in a differential drive configuration, a microcontroller to control movement and an ultrasonic sensor to detect obstacles. The robot is built to wander around without bumping into objects. Explain how stubs could be used in testing to take the place of incomplete subsystems. Be specific.

7.4   Consider that you have an op amp integrated circuit package, such as the LM741 in Appendix C. What type of testing would be appropriate for testing this device? Write a short test plan for doing so.

7.5   Explain under what situations a matrix test is appropriate.

7.6   Explain under what situations a step-by-step is test appropriate.

7.7   Consider the stages of unit testing, integration testing, and acceptance testing. For each of these stages, identify the corresponding requirements that each test should be traceable to.

7.8   **Project Application.** Develop a unit test for your project. The unit tests should apply to the lowest level units in the design.

7.9   **Project Application.** Develop an integration test for your project. The integration tests should apply to the higher levels of the design architecture.

7.10   **Project Application.** Develop an acceptance test for your project. The acceptance tests should apply to the engineering requirements developed for the system.

Chapter 8   System Reliability

Quality is never an accident. It is always the result of intelligent effort. —John Ruskin

A typical design project in your academic career may never leave the confines of a laboratory. However, in industry, engineers develop systems that are used by the public at large, and issues beyond the functionality, such as reliability, safety, and maintainability become important factors in the success of the design. Over the past 20 years, industry has made a great shift to address reliability through the adoption of processes such as Quality Functional Deployment (QFD), Six-Sigma, and Robust Design. While other chapters have addressed some elements of these processes, the objective of this chapter is to examine system reliability. Reliability attempts to answer the question of how long a system will operate without failing. Answering this question has inherent uncertainty and requires the use of probability and statistics. This chapter presents a review of basic probability theory and applies it to estimate the behavior of real-world devices. Reliability at the component and system levels is considered.

Learning Objectives

By the end of this chapter, the reader should:

- Have a familiarity with the basic principles of probability and understand how they apply to reliability theory.
- Understand the mathematical definition and meaning of failure rate, reliability, and mean time to failure.
- Understand how to determine the reliability of a component.
- Understand how to derate the power of electronic components for use under different operating temperatures.
- Understand how to determine the reliability of different system configurations.

DILBERT® by Scott Adams

Figure 8.1 Dogbert's Six Sigma Program. (Dilbert © United Feature Syndicate. Reprinted by permission.)

8.1 Probability Theory Review

Probability theory provides a formal framework to study chance events. It is a powerful tool for modeling engineering systems and is a requisite for reliability estimation. Although this section provides a review of some important concepts from probability, it is assumed that the reader is versed in the basics of probability theory.

In order to apply probability, some general definitions are examined first. An *experiment* is the process of measuring or quantifying the state of the world. The particular outcome of an experiment is an *event* (e_i), while the *event space* (E) is the set of all possible outcomes of the experiment. For example, consider an experiment where a six-sided die is rolled. The experiment is rolling the die and observing the outcome, the event is the particular outcome observed, and the event space for the experiment is the set $E = \{1,2,3,4,5,6\}$. The outcomes do not have to be numerical values. Another example experiment is tossing a coin, in which case the event space is $E = \{Heads, Tails\}$. Both are examples of a discrete event space because there are a finite number of experimental outcomes. In a discrete event space, the union of all the possible experimental outcomes defines the event space. If e_i is the ith event in a discrete event space, then the event space is given by the union

$$E = \bigcup e_i. \tag{1}$$

The probability of an event indicates how likely it is for an event to occur. This is quantified by the probability operator, $P()$, that assigns to each event a real number between 0 and 1. The probability is the percentage of times that an event would occur if the experiment were repeated an infinite number of times (the Law of Large Numbers). Two of the three fundamental axioms on which probability theory is built are

$$P(e_i) \geq 0 \tag{2}$$

$$P(E) = 1. \tag{3}$$

The first axiom indicates that all probabilities are non-negative, while the second is a restatement of the event space definition—the outcome of an experiment must be an element of the event space. Armed with these definitions and axioms, some important concepts from probability are now examined.

8.1.1 Probability Density Functions

Not all event spaces are discrete as in the case of rolling a die or flipping a coin. Consider an experiment where the objective is to measure temperature. Clearly, such a measurement requires a variable having a continuous range of possible values. A random variable is defined as the outcome of an experiment that has a continuum of possible values. Random variables have a mathematical function known as the *probability density function* (PDF) associated with them, which when integrated, yields the probability of a range of events. A PDF is typically denoted as $p_X(x)$, where X takes values over the event space. Standard notation identifies random variables using upper case variables as the subscript for the PDF. The variable inside the parentheses is a lowercase dummy variable that does not have to match the random variable, but typically does. A question that the PDF allows us to ask is *"What is the probability that a random variable is in some range?"* Consider the case where the objective is to determine the probability that the random variable X lies between two values a and b. Written using the probability operator, this is indicated as $P(a \le X \le b)$. It is determined from the PDF as follows

$$P(a \le X \le b) = \int_a^b p_X(x)dx. \tag{4}$$

Conceptually, this probability represents the area under the PDF between the two limits of integration as shown in Figure 8.2.

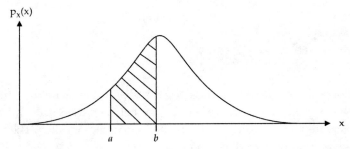

Figure 8.2 A probability density function. The area under the curve represents the probability that the random variable X lies in the interval [a,b].

Let's examine a few more important properties of probability density functions. The first, which is analogous to (3), indicates that the probability of the event space occurring is equal to one. This is known as the normalization property and it is expressed as

$$\int_{-\infty}^{\infty} p_X(x)dx = 1.\tag{5}$$

Another interesting result is obtained by trying to determine the probability that a random variable takes on an exact value, for example $P(X = a)$. That is determined from the integral

$$P(X = a) = \int_a^a p_X(x)dx = 0.\tag{6}$$

This is a somewhat counterintuitive result—it indicates that the probability a random variable can take on a particular value is zero. Does this make any sense? Consider an experiment where the objective is to measure a voltage value for a random variable V. Now consider the question, "*What is the probability that the result of a voltage measurement equals π (the irrational number) volts?*" In practice, this question is impossible to answer because the precision required of the meter is infinite and contrary to its construction. So the mathematical and practical results are in harmony. There is a way around this dilemma, which is to determine the probability that the random variable is within a small range about the target value as follows

$$P(\pi < V < \pi + \Delta v) = \int_{\pi}^{\pi + \Delta v} p_V(v)dv \approx p_V(\pi)\Delta v.\tag{7}$$

This means that the probability a random variable is within a small range about a given value is approximated by the product of the PDF evaluated at the value and the size of the range.

8.1.2 Mean and Variance

Two useful and well-known statistics that are determined from the PDF are the mean (μ) and variance (σ^2). They are found from the PDF as follows

$$\mu_X = \int_{-\infty}^{\infty} x p_X(x)dx\tag{8}$$

$$\sigma_X^2 = \int_{-\infty}^{\infty} (x - \mu)^2 p_X(x)dx.\tag{9}$$

The *mean* is analogous to the center of mass of the PDF; it is also known as the average value. The *variance* is the average of the squared difference between the mean and the values of the PDF, where the squared term ensures that a positive difference is taken. The square root of the variance is known as the standard deviation (σ).

8.1.3 Common Probability Density Functions

There are many PDFs available for describing the seemingly random variations in the behavior of observed systems and phenomena. In this section, three common PDFs (normal, exponential, and uniform) are presented.

The Normal Density

The most common density function encountered in the physical sciences and engineering is the normal density. Many population variations can be described by a normal density. For example, the resistance values of a large batch of 2.2k ohm resistors would likely follow a normal density. The normal density is defined as

$$p_X(x) = \frac{1}{\sqrt{2\pi}\sigma} e^{-\frac{1}{2}\left(\frac{x-\mu}{\sigma}\right)^2} .$$ (10)

The mean, μ, and standard deviation, σ, are part of the definition of the PDF and used to alter the shape of the density to suit the particular need. The normal PDF is plotted in Figure 8.3. Varying μ allows the overall function to be shifted along the x-axis, while increasing σ spreads (or flattens) the function out. Calculating probabilities from the normal density can be done (although it takes a bit of work mathematically) so they are usually computed from something known as the Cumulative Distribution Function that is presented shortly.

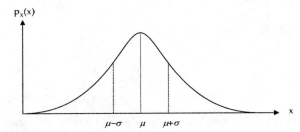

Figure 8.3 A normal density function with the mean (μ) and standard deviation (σ) shown.

The Uniform Density

The uniform density, plotted in Figure 8.4, models the outcome of an experiment where all outcomes are equally likely. Mathematically, the PDF for a uniform density is given by

$$p_X(x) = \frac{1}{b-a}, \quad a \le x \le b,$$ (11)

where a and b are selected to meet the demands of a particular problem.

Figure 8.4 The uniform density on the interval $[a,b]$.

The Exponential Density

Exponential densities are often utilized to model time dependent functions, such as inter-arrival times between data packets in communication systems. As shown later, the exponential density also describes the behavior of component failures as a function of time. The mathematical description of an exponential density is

$$p_X(x) = \lambda e^{-\lambda x}, \quad x \geq 0, \lambda \geq 0 . \tag{12}$$

The PDF is characterized by the parameter λ, which affects the shape of the curve as demonstrated in Figure 8.5.

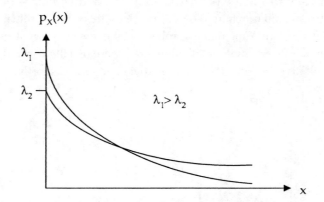

Figure 8.5 The exponential density for two different λ values.

8.1.4 Cumulative Distribution Functions

An important class of questions can be phrased as, "*What is the probability that a random variable X is less than value a?*" For example, the objective might be to determine the probability that an electronic component will malfunction within two years. Returning to the first question, it is clear that the goal is to determine the probability $P(X < a)$, which is found by integrating the PDF. This result is generalized by allowing the upper limit of integration to take on an arbitrary value that spans the range of the random variable. This produces a new function, known as the *Cumulative Distribution Function* (CDF), which is the integral function of the PDF and is defined as

$$CDF(x) \equiv \int_{-\infty}^{x} p_X(y) dy . \tag{13}$$

8.2 Reliability Prediction

Our main interest in the study of probability stems from the desire to quantify the reliability of a system. The following is a formal mathematical definition of *reliability*.

> **Definition**: Reliability, R(t), is the probability that a device is functioning properly (has not failed) at time *t*.

In order to determine R(t), it is necessary to first introduce some related mathematical entities and their meanings. The ***failure rate***, $\lambda(t)$, of a device is the expected number of failures per unit time. The failure rate is measured by operating a batch of devices for a given time interval and noting how many fail during that interval. A typical graph of failure rate versus time has the bathtub shape shown in Figure 8.6. The high initial failure rate is a result of manufacturing defects often referred to as infant mortality. Consequently, many manufactures will "burn-in" devices at the factory, so that if they fail, they do so before being sold. After the infant mortality phase, devices enter a phase of constant failure rate, where $\lambda(t)= \lambda$, known as the service life. Estimates for λ are determined empirically by testing a large number of components. They are usually expressed as a unit failure per a given number of hours, for example $\lambda = 1\,\text{failure}/10^6\,\text{hours}$. After some period of time, devices start to wear-out and the failure rate increases. This usually happens as a result of mechanical wearing with age and use. Properly designed electronic devices will not have a wear-out region, instead continuing on at a constant failure rate. This applies only to the electronic devices themselves, not necessarily to complete systems that will likely contain mechanical devices.

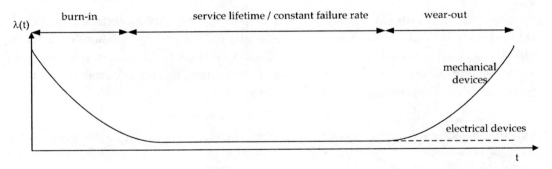

Figure 8.6 Failure rate as a function of time, also known as the bathtub curve.

In addition to failure rate, a PDF for the *failure time* of the device, $f_T(t)$, is defined, where the random variable is time *T*. This function allows the question to be asked "*What is the probability that a device will fail between time t₁ and t₂?*" It is important to note the difference between $\lambda(t)$ and $f_T(t)$. The failure rate tells us the average rate that a collection of identical devices will fail at a given time *t*, while $f_T(t)$ is a PDF used to determine the probability that a given device will fail within a specified time period. A CDF for $f_T(t)$ is determined as

$$F(t) = \int_0^t f_T(\tau)d\tau . \tag{14}$$

$F(t)$ answers the question "*What is the probability that the device has failed by time t?*" and it is also known as the **failure function**. Take a few seconds to go back and review the definition of $R(t)$. It is clear that $R(t)$ is directly related to $F(t)$ and is its complement. The relationship between the two is

$$R(t) = 1 - F(t) . \tag{15}$$

Since $F(t)$ is a CDF, it increases monotonically from an initial value of 0 to a maximum value of 1 as time goes to ∞ as shown in Figure 8.7. Conversely, $R(t)$ starts at a value of 1 at time zero and decreases monotonically to a value of 0.

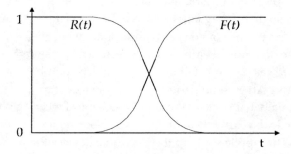

Figure 8.7 Example reliability and failure functions.

Since $\lambda(t)$ represents data that is measured empirically, it is useful to establish a relationship between $\lambda(t)$ and the ultimate goal of reliability, $R(t)$. To do so, a relationship between $\lambda(t)$, $R(t)$, and $f_T(t)$ is established as follows. Consider a small period of time between t and $t + \Delta t$, and determine the probability of device failure during this period. From the approximation developed in (7), this probability is given by

$$P(\text{failure between } t \text{ and } \Delta t) \approx f_T(t)\Delta t . \tag{16}$$

How is this probability related to $R(t)$ and $\lambda(t)$? $R(t)$ provides the probability that the device is working at time t and $\lambda(t)$ gives the probability that the device will fail at time t. The product of $R(t)$, $\lambda(t)$, and Δt gives the same probability of failure in (16)

$$P(\text{failure between } t \text{ and } \Delta t) = R(t)\lambda(t)\Delta t . \tag{17}$$

Equating (16) and (17) provides the desired relationship between the three quantities

$$f_T(t) = R(t)\lambda(t) , \tag{18}$$

that is fundamental in establishing the connection between $R(t)$ and $\lambda(t)$. However, the PDF $f_T(t)$ needs to be eliminated from (18). This is accomplished through its relationship to the CDF $F(t)$, and thus $R(t)$, as follows

$$f_T(t) = \frac{d}{dt}F(t) = \frac{d}{dt}[1 - R(t)] = -\frac{d}{dt}R(t). \tag{19}$$

Equating this result with (18) produces

$$-\frac{d}{dt}R(t) = R(t)\lambda(t) \implies \frac{-\frac{d}{dt}R(t)}{R(t)} = \lambda(t). \tag{20}$$

Integrating both sides gives

$$\int_0^t \left(\frac{-\frac{d}{d\tau}R(\tau)}{R(\tau)} \right) d\tau = \int_0^t \lambda(\tau)d\tau \implies \ln(R(t)) = \int_0^t \lambda(\tau)d\tau \tag{21}$$

and solving for $R(t)$ produces the final result for reliability as a function of $\lambda(t)$

$$\boxed{R(t) = \exp\left(-\int_0^t \lambda(\tau)d\tau \right).} \tag{22}$$

During the service lifetime phase, the failure rate is constant, which simplifies to

$$\boxed{R(t) = \exp(-\lambda t).} \tag{23}$$

This important result is now applied in Example 8.1.

Example 8.1 Transistor reliability.

Problem: Consider a transistor with a constant failure rate of $\lambda = 1/10^6$ hours . What is the probability that the transistor will be operable in 5 years?

Solution: This solution is found using the reliability function for a constant failure rate in (23) as follows.

$$R(t) = \exp(-\lambda t)$$

$$R(5 \text{ years}) = \exp\left(-\frac{1}{10^6 \text{ hours}} \times \frac{24 \text{ hours}}{\text{day}} \times \frac{365 \text{ days}}{\text{year}} \times 5 \text{years} \right)$$

$$= \exp(-0.0438)$$

$$= 0.957$$

$$= \underline{95.7\%}.$$

8.2.1 Mean Time to Failure

The *mean time to failure* (MTTF) is a quantity which answers the question, "*On average how long does it take for a device to fail?*" From its definition, it is apparent that the MTTF is the mean value of the random variable T (failure time). It is determined from the PDF and the definition of the mean in (8) as follows

$$MTTF = \int_0^\infty t f_T(t) dt. \tag{24}$$

At this point the form of the PDF for $f_T(t)$ is not known, but it can be found from (19) since it is the negative derivative of $R(t)$. Assuming the form of $R(t)$ found in (23) for a constant failure rate gives

$$f_T(t) = -\frac{d}{dt} R(t) = \lambda e^{-\lambda t}. \tag{25}$$

This means that under the condition of a constant failure rate, the failure PDF follows an exponential density. The MTTF is found from $f_T(t)$ via integration by parts to be

$$\boxed{MTTF = \int_0^\infty t e^{-\lambda t} dt = \frac{1}{\lambda}.} \tag{26}$$

This makes intuitive sense because λ is the expected number of failures per unit time for a device. Consequently, the reciprocal of λ is the expected time between failures or MTTF. Let's consider a few examples.

Example 8.2 Transistor MTTF.

Problem: Consider the transistor in Example 8.1. (a) Determine the MTTF, and (b) the reliability at the MTTF.
Solution:

(a) From (26) the MTTF $= \dfrac{1}{\lambda} = \dfrac{1}{(1/10^6 \text{ hours})} = 10^6$ hours

$\qquad\qquad = \underline{114 \text{ years}}$.

(b) From (23) the reliability at 114 years is

$$R(t) = \exp(-\lambda t)$$

$$R(114 \text{ years}) = \exp\left(-\frac{10^6 \text{ hours}}{10^6 \text{ hours}}\right) = \exp(-1) = 0.368$$

$$= \underline{36.8\%}.$$

This is a bit counterintuitive. Although the average time between transistor failures is 114 years, an individual transistor has only a 36.8% chance of surviving to 114 years. It would seem logical that the reliability at 114 years should be 50% and that the transistor would have a 50-50 chance of failing. This would be true if $f_T(t)$ were symmetric about its mean, but that is not the case for the exponential density.

Example 8.3 Human lifespan estimation.

Problem: Data shows that for a 30 year old population, the failure (death) rate is constant with approximately 1.1 deaths per 1000 people per year. Given this data, estimate the MTTF of humans.

Solution: In order to find MTTF, λ is needed. From the information given it is

$$\lambda = \frac{(1.1/1000)\,\text{failures}}{1\,\text{year}} = \frac{1.1\,\text{failure}}{10^3\,\text{years}} = \frac{1\,\text{failure}}{909\,\text{years}}$$

From this MTTF is computed as

$$\text{MTTF} = \frac{1}{\lambda} = \underline{909\ \text{years}}!$$

While great news for those of us seeking longevity, this calculation is clearly wrong since the upper limit on human lifespan is empirically known to be about 120 years. Why is this so? Serious problems arise if $R(t)$ is used in situations where the underlying assumption is invalid. The results in (23) and (26) apply only if the failure rate is constant. Although that is nearly true for people in their 20s and 30s, it is not true as people age. People do wear out and the failure rate increases with age.

8.2.2 Failure Rate Estimates

The overriding objective of this chapter is to estimate the future behavior of devices that are used in electrical and computer systems. The particular behavior of interest is the state of a device's functionality—the reliability, is it working or has it failed? Equation (23) indicates that it is fairly straightforward to determine reliability, if the failure rate (λ) is known and is constant. One question to consider is what factors influence the failure rate of a device. Many of us probably have had experiences in the laboratory where we have caused devices to fail by subjecting them to conditions outside of the normal operating bounds, notably excessive current, power, or heat. In those cases the devices probably failed, or burned up, due to the operating conditions outside of the allowed bounds for the device. However, even when operated within the allowable norms of a device's operating conditions, variations in factors such as power, operating voltages, and temperatures impact λ.

The United States Military has kept copious records of device failures in the field and the conditions under which they operated. These records are synthesized in a handbook entitled

Reliability Prediction of Electronic Equipment [MIL-HDBK-217F] that provides failure rates for various analog and digital components, along with adjustment factors to account for operating conditions, the environment, and device quality. Categories of devices included in the handbook are switches, fuses, diodes, optoelectronic devices, and microelectronic devices (op amps, logic devices, microcontrollers, microprocessors). The handbook was last published in 1991 and has been discontinued, but it is still widely accepted and used. Bellcore (subsequently Telcordia) has developed newer models [Tel96] based upon MIL-HDBK-217F that were updated to better predict the reliability of components. MIL-HDBK-217F is used here since it is freely available in the public domain.

Failure rates for resistors, capacitors, transistors, and integrated circuits from MIL-HDBK-217F are included in Appendix B. For each device, a base failure rate is given, λ_b, and multiplied by a number of adjustment factors, denoted by the symbol π, to estimate the device failure rate λ. Each adjustment factor has a unique subscript and their values are found from tables or equations in the handbook.

For example, consider the low frequency field effect transistor on page 260 in Appendix B. The overall failure rate is given by the equation $\lambda = \lambda_b \pi_T \pi_A \pi_Q \pi_E$ failures$/10^6$ hours. λ_b is the base failure rate that is directly read from a table, π_T is a temperature factor which is computed from an exponential equation (be careful to use the junction temperature as indicated on page 254 in the appendix), π_A is an application factor that depends upon how the device will be used, π_Q is a quality factor, and π_E is an environmental factor. The quality factor table lists some strange names and values from 0.7 to 8.0. The quality factor describes the level of burn-in and screening each device receives before leaving the factory. Joint Army/Navy (JAN, JANTX, and JANTXV) quality factors are the highest standard, and are usually required only for space vehicles. That individual attention to burn-in means that JAN parts are expensive and most JAN devices have passed their infant mortality phase before leaving the factory. When determining failure rate for a device from a table, it is common practice to always round parameters or values pessimistically so that the evaluation is a worst case analysis of its performance. That way the device should perform with a higher reliability when embedded into a system, hopefully causing only pleasant surprises in operation. Finally, the factor π_E is based upon the different operating environments that are identified on pages 249 and 250 in Appendix B. Example 8.4 demonstrates the application of the MIL-HDBK-217F standard for reliability estimation.

In summary, it is possible to estimate the reliability of devices if the failure rate is known and it is constant. The US Military and Telecordia handbooks provide guidance for estimating failure rates, and thus the component reliability. It must be kept in mind that they are estimates and not guaranteed to predict the exact performance. It is also apparent that this can become a rather time-consuming process if there are many components in a system, and thus the use of reliability software packages may be warranted.

Example 8.4 Reliability estimation using the MIL-HDBK 217F.

Problem: Consider the circuit below that contains a bipolar junction transistor (BJT). This electronic circuit is a simple digital logic inverter. When the input voltage V_I is 0, the BJT is off, no current flows through any branches of the device, and the output voltage, V_O, is 5V. When the input is 5V (high) the BJT goes into saturation due to the high base current (low R_B), producing a 50mA collector current, a large voltage drop across R_C, and an output voltage close to 0V. The average collector current for the two states is 25mA, producing an average power of 125mW (25mA×5V). Assume that the circuit is used in a missile launcher, the ambient temperature is 25°C, and that JANTX quality parts are used. Determine the MTTF and reliability for the 2N3904 BJT (a low power, low frequency BJT) in 20 years.

Solution: The objective of this problem is to determine the failure rate, from which the MTTF and reliability are estimated. From the MIL-HDBK-217F data in Appendix B, the failure rate is

$$\lambda = \lambda_b \pi_T \pi_A \pi_R \pi_S \pi_Q \pi_E \frac{\text{failures}}{10^6 \text{ hours}}$$

The base failure rate is given directly as $\lambda_b = 0.00074$. π_T is the temperature factor and its value is determined from the relationship

$$\pi_T = \exp\left(-2114\left(\frac{1}{T_J + 273} - \frac{1}{298}\right)\right).$$

where T_J is the junction temperature. As indicated in Appendix B, it is computed as

$$T_j = T_A + \theta_{JA} P_D$$

$$= 25°C + \left(200\frac{°C}{W}\right)(125x10^{-3}W) = 50°C.$$

The thermal resistance, θ_{JA}, is read from the 2N3904 datasheet (pages 267 and 268 in Appendix C) and will be examined in more detail shortly. The temperature factor is

$$\pi_T = \exp\left(-2114\left(\frac{1}{50 + 273} - \frac{1}{298}\right)\right) = 1.73.$$

π_A is an application factor (switched or linear amplification), and the value for the switched logic inverter is $\pi_A = 0.70$. π_R is a power rating factor that is computed based upon the maximum rated power dissipation of the 2N3904 (625mW from the component datasheet in Appendix C) as follows

$$\pi_R = (P_R)^{0.37} = (0.625)^{0.37} = 0.84 .$$

π_S is a stress factor that is computed from the ratio of the maximum collector-emitter voltage over the maximum rated value of the device. In this circuit, the maximum value of V_{CE} is 5V (when the device is off and no current flows), while the maximum rated value of V_{CE} from the 2N3904 datasheet is 40V.

$$V_S = \frac{\text{applied } V_{CE}}{\text{rated } V_{CE}} = \frac{5}{40} = 0.125$$

$$\pi_{s'} = 0.045 \exp(3.1 \times V_s) = 0.045 \exp(3.1 \times 0.125) = 0.066.$$

π_Q is the quality factor, and based upon the fact that a JANTX part is used $\pi_Q = 1.0$. π_E is the environmental factor, and for the missile launch application, $\pi_E = 32.0$.

All of this is brought together to compute the failure rate. The product of the adjustment factors is computed as

$$\lambda = \lambda_b \pi_T \pi_A \pi_R \pi_S \pi_Q \pi_E \frac{\text{failures}}{10^6 \text{ hours}}$$

$$= \left(\frac{7.4x10^{-4} \text{ failures}}{10^6 \text{ hours}} \right)(1.73)(0.7)(0.84)(0.066)(1.0)(32.0)$$

$$= \frac{1.59x10^{-3} \text{ failures}}{10^6 \text{ hours}}$$

This allows estimation of the MTTF and requested reliability

$$\text{MTTF} = \frac{1}{\lambda} = 8.6x10^8 \text{ hours } = \underline{71,804 \text{ years.}}$$

$$R(20 \text{ years}) = \exp\left(-\frac{20 \text{ years}}{71,804 \text{ years}} \right) = 0.9997$$

$$= \underline{99.97\%}.$$

In conclusion, the BJT is estimated to be highly reliable in 20 years.

8.2.3 Thermal Management and Power Derating

One of the quantities computed in Example 8.4 was the junction temperature, T_J, which was computed from the power dissipated in the device and a quantity known as thermal resistance, θ. It is important to understand this in more detail, since it impacts the reliability of microelectronic devices. Furthermore, if the junction temperature exceeds a certain value, the device will fail. Thus, thermal management issues need to be taken into account. We start with a physical model in Figure 8.8 (a), which has a junction (the integrated circuit or device), enclosed by a case (the packaging of the device), surrounded by ambient environmental conditions. In part (b) a heat sink is included which aids in thermal transfer. Each element has associated with it a quantity known as thermal resistance that measures the ability of that particular element to transfer heat to another element. A result from heat transfer for electronics is that changes in temperature (ΔT) are proportional to the product of power dissipation (P_D) and the thermal resistance. This relationship is

$$\Delta T = P_D \theta \, . \tag{27}$$

It is similar to Ohm's Law where the change in temperature, power, and thermal resistance (units = °C/W) are analogous to voltage, current, and electrical resistance respectively. The total thermal resistance between two elements, such as between ambient and the junction, is the sum of all thermal resistances between them. In the case with no heat sink, this produces a junction to ambient resistance of $\theta_{JA} = \theta_{JC} + \theta_{CA}$, while in the case with a heat sink, $\theta_{JA} = \theta_{JC} + \theta_{CS} + \theta_{SA}$. In the case of the heat sink, θ_{CA} is replaced by the sum $\theta_{CS} + \theta_{SA}$, which has a lower combined thermal resistance and greater ability to dissipate heat.

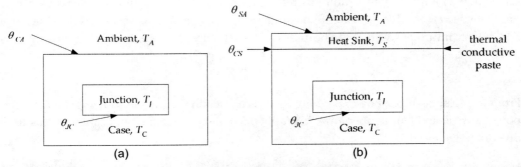

Figure 8.8 Physical model of microelectronic devices with thermal junctions and temperatures indicated. (a) Device inside casing. (b) Device inside casing with a heat sink added.

This Ohm's Law type of relationship means that thermal transfer can be modeled using familiar resistive circuits as shown in Figure 8.9. Based upon this circuit model, the temperature can be found at different points from the thermal resistance and power dissipation. Most importantly the junction temperature is found as

$$T_J = T_A + P_D \theta_{JA} \, . \tag{28}$$

Figure 8.9 Resistive models for thermal transfer in microelectronic devices. (a) Device with no heat sink. (b) Device with a heat sink added.

Let's now apply these results. Manufacturer datasheets typically identify the absolute maximum power dissipation and note that the device should be derated if operated at ambient conditions above room temperature ($T_A = 25°C$). The datasheets also supply a maximum junction temperature for the device. It is clear from the resistive model that, for a fixed power dissipation, the junction temperature increases along with ambient temperature. If the maximum junction temperature is exceeded, the device will be destroyed. Another way to look at this is that as ambient temperature increases, the maximum amount of power a device can dissipate decreases. This decrease in maximum power dissipation is known as *derating*. From (28) the maximum power that can be dissipated in a device at a given ambient temperature is

$$P_{D,\max} = \frac{T_{J,\max} - T_A}{\theta_{JA}} . \qquad (29)$$

From this relationship, a power derating curve is plotted in Figure 8.10 showing the maximum power versus ambient temperature. Example 8.5 demonstrates the application of this to the inverter in Example 8.4.

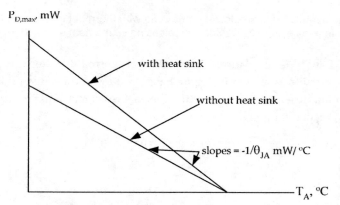

Figure 8.10 Typical power derating curves.

Example 8.5 Power Derating for the Inverter Circuit.

Problem: Assume the circuit in Example 8.4 is operating at an ambient temperature $T_A = 120°C$ and that no heat sink is used. (a) Determine the derated power and if the design is within the manufacturer's limits for power dissipation at this temperature and, (b) re-compute the reliability at 20 years based upon this elevated operating temperature.

Solution:
(a) From the manufacturer datasheet in Appendix C, the 2N3904 BJT has a thermal resistance of $\theta_{JA} = 200°\,C/W$ and a maximum junction temperature of 150°C. From this the maximum power is computed from (29) as

$$P_{D,\,max} = \frac{(150 - 120)°C}{200°C/W} = 150mW \ .$$

The derated, or maximum, power at this temperature is 150mW. Clearly, the 125mW of power dissipated as determined in Example 8.4 for the BJT is within this derated limit.

(b) To compute the failure rate the junction temperature and π_T are re-computed.

$$T_J = T_A + P_D\theta_{JA} = 120°C + (125 \times 10^{-3}W)\left(\frac{200°C}{W}\right) = 145°C$$

$$\pi_T = \exp\left(-2114\left(\frac{1}{145 + 273} - \frac{1}{298}\right)\right) = 7.66.$$

With this new value, the value of $\lambda = 7.00 \times 10^{-3}/10^6$ hours, and the reliability is reduced slightly to 99.88%. Note, however, the junction temperature is quite high at 145°C and further increases in temperature would likely destroy the device.

8.2.4 Limits of Reliability Estimation

It must be kept in mind that the reliability estimates are just that, estimates, and there are limitations in their use. First, realize that the failure rate data comes from accelerated stress tests, where devices are put under stress beyond normal operating conditions, and from these the failure rates are estimated. (Nobody sits around waiting 20 years for the devices to fail!) The tests are based upon mathematical models for the failure rate and the device lifetime. Secondly, there are other factors that influence reliability that are not addressed by λ, such as the manufacturing processes used, the quality of manufacturing technologies, shock, and corrosion. Part of the value of reliability estimation is for comparative purposes when evaluating different design options. Applying these methods forces the designer to consider the operating conditions and factor them into the design.

8.3 System Reliability

The previous section focused on determining the reliability of a single device. It is natural to ask, *"How can the reliability of a system consisting of many devices be determined?"* In order to derive the overall reliability of a multi-component system, it is necessary to take into account the overall system structure.

8.3.1 Series Systems

Consider the inverter circuit in Example 8.4—failure of any one component in the circuit would lead to the failure of the overall system or circuit. Conceptually, a system in which the failure of a single component (or subsystem) leads to failure of the overall system is known as a *series system*. Figure 8.11 shows a block diagram of a series system composed of boxes S_1, S_2, ..., S_n that represent the components, or the subsystems, of a larger system.

Figure 8.11 A series system consisting of components, or subsystems S_1, S_2, ..., S_n.

To compute the overall reliability of a series system, $R_s(t)$, it is assumed that the failure of subsystems or components are independent events. The system is operable only if subsystems S_1 and S_2 ... and S_n are all simultaneously operating. Therefore, the probability of the overall system operating is given by the product of reliabilities for all of the subsystems as follows

$$R_s(t) = R_1(t)R_2(t)\cdots R_n(t) = \prod_{i=1}^{n} R_i(t).$$ (30)

It is important to remember that failures are assumed to be independent events, just as flipping a coin twice is considered two independent events. The overall system reliability is less than or equal to that of any single subsystem, since all reliability values are ≤ 1. Thus R_s decreases as the number of subsystems increases. Assuming a constant failure rate for all system components gives the following result for the overall system reliability

$$R_s(t) = e^{-\lambda_1 t} e^{-\lambda_2 t} \cdots e^{-\lambda_n t} = \exp\left(-\sum_{i=1}^{n} \lambda_i t\right).$$ (31)

This leads to a series system failure rate and MTTF of

$$\lambda_s = \sum_{i=1}^{n} \lambda_i \quad \text{and} \quad \text{MTTF}_s = \frac{1}{\lambda_s}.$$ (32)

Example 8.6 revisits the inverter problem where the failure rates of all components are considered for system reliability estimation.

Example 8.6 Inverter circuit reliability.

Problem: For the system in Example 8.4 estimate (a) the overall system reliability in 20 years, and (b) the MTTF. Assume room temperature and that ¼ watt fixed composition resistors are used.

Solution:

(a) Conceptually this is a series system—if any of the individual components fail, then the overall system will fail. That means that failure rates for the two resistors are needed in addition to the value previously computed for the transistor. They depend upon the power dissipated in each resistor, which is 125mW and 0.9mW for the collector and base resistors respectively. The failure rate for a fixed composition resistor from MIL-HDBK-217F is

$$\lambda_{resistor} = \lambda_b \pi_R \pi_Q \pi_E \text{ failures}/10^6 \text{ hours}$$

For the collector resistor, Rc, the base failure rate is computed as

$$\lambda_b = 4.5 \times 10^{-9} \exp\left(12\left(\frac{T+273}{343}\right)\right)\exp\left(\frac{S}{0.6}\left(\frac{T+273}{273}\right)\right)$$

$$= 4.5 \times 10^{-9} \exp\left(12\left(\frac{25+273}{343}\right)\right)\exp\left(\frac{.125/.25}{0.6}\left(\frac{25+273}{273}\right)\right)$$

$$= 3.77 \times 10^{-4}.$$

The S term is the ratio of power dissipated to the maximum power rating. The values $\pi_R = 1.0$, $\pi_Q = 15.0$, and $\pi_E = 27.0$ are directly read from tables. Thus the overall failure rate for the collector resistor is

$$\lambda_{resistor1} = (3.77 \times 10^{-4})(1.0)(15.0)(27.0) = 1.53 \times 10^{-1} \text{ failures}/10^6 \text{ hours}.$$

The process for the base resistor, RB, is similar, and results in

$$\lambda_{resistor2} = 6.1 \times 10^{-2} \text{ failures}/10^6 \text{ hours}.$$

The total failure rate is given from (32) as

$$\lambda_s = \lambda_{BJT} + \lambda_{resistor1} + \lambda_{resistor2} = 0.215 \text{ failures}/10^6 \text{hours}$$

$$R_S(t) = \exp(-\lambda_s t) = \exp\left(-\frac{0.215}{10^6 \text{ hours}} \times \frac{24 \text{ hours}}{\text{day}} \times \frac{365 \text{ days}}{\text{year}} \times 20 \text{ years}\right) = 0.963$$

$$= \underline{96.3\%}.$$

Since resistors are pretty reliable devices, the overall system reliability decreases only a small amount relative to that of the BJT itself.

(b) The MTTF is given by $1/\lambda_s$ which in this case is <u>531 years</u>.

8.3.2 Parallel Systems

From (30) it is clear that as more components are added to a series system, the reliability decreases. It is natural to ask if the reliability can be increased. The use of redundancy gives us a method to answer in the affirmative. A design has **redundancy** if it contains multiple modules performing the same function where a single module would suffice. By its very nature redundancy allows improperly functioning modules to be switched out of the system without affecting its behavior. With redundancy the overall system functions correctly when any one of the submodules is functioning. Figure 8.12 shows a simplified view of a **parallel system** with subsystems S_1, S_2, ..., S_n.

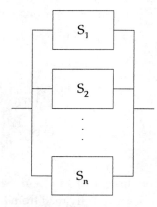

Figure 8.12 A parallel, or redundant, system consisting of subsystems S_1, S_2, ..., S_n

In order to compute the reliability of a parallel system, note that a parallel system functions correctly when S_1 is functioning correctly, or S_2 is functioning correctly, ... or S_n is functioning correctly. It would be nice if it were possible to write an equation stating that $R_s(t) = R_1(t) + R_2(t) + \cdots + R_n(t)$, where + is the logical OR operator. Unfortunately, there is no direct way to realize the OR operation in probability theory. This is resolved by working with failure function, $F(t)$, instead. The probability that the system will fail by time t, $F_s(t)$, is equal to the probability that subsystem S_1 will fail and S_2 will fail and ... S_n will fail. This probability is expressed mathematically as

$$F_S(t) = F_1(t)F_2(t)...F_n(t) = \prod_{i=1}^{n} F_i(t) = \prod_{i=1}^{n}\left[1 - R_i(t)\right]$$

(33)

The overall system reliability of the parallel system is found from this as

$$\boxed{R_S(t) = 1 - F_s(t) = 1 - \prod_{i=1}^{n}\left[1 - R_i(t)\right]}$$

(34)

As more redundant components are added to a parallel system, additional $1 - R_i(t)$ terms are introduced into the product term. This decreases the value of the product, hence the overall system reliability is increased as more redundant systems are added.

In order for a parallel system to work, a mechanism must be in place to monitor each of the subsystems to make sure that they are operating correctly. Developing circuits to detect failures and control the switching between subsystems can be complex and is not considered here. Special care must be paid to the switching circuit itself, as malfunction of this circuit could lead to an overall system failure. An example of parallel system reliability is given in Example 8.7.

Example 8.7 Reliability of a Redundant Array of Independent Disks (RAID).

Problem: In a RAID, multiple hard drives are used to store the same data, thus achieving re- dundancy and increased reliability. One or more of the disks in the system can fail and the data can still be recovered. However, if all disks fail, then the data is lost. For this problem, assume that the individual disk drives have a failure rate of $\lambda = 10\,\text{failures}/10^6\,\text{hours}$. How many disks must the system have to achieve a reliability of 98% in 10 years?

Solution: The reliability of a parallel system with redundancy is given by (34). Since all of the disks are identical, the expression simplifies to

$$R_S(t) = 1 - \left[1 - R_i(t)\right]^n$$

$$0.98 \leq 1 - \left[1 - \exp\left(-\frac{10}{10^6\,\text{hours}} \times \frac{24\,\text{hours}}{\text{day}} \times \frac{365\,\text{days}}{\text{year}} \times 10\,\text{years}\right)\right]^n$$

$$0.98 \leq 1 - \left[1 - 0.42\right]^n$$

$$\Downarrow$$

$$0.02 \leq (0.58)^n$$

$$\Downarrow \qquad\qquad \text{(take logarithm of both sides)}$$

$$\log(0.02) \leq n\log(0.58)$$

$$\Downarrow$$

$$n \geq 7.2$$

In order to achieve this reliability, _n=8_ disks are required. The reliability of each individual disk is low at 42%, but using redundancy, the overall system reliability is quite high.

8.3.3 Combination Systems

Many real systems do not fit neatly into either parallel or series reliability models as shown in Figure 8.13. Rather, they may be a combination of the two, and such systems will be referred to as combination systems. One way to determine the reliability of a combination system is to utilize the results obtained for series and parallel systems in (30) and (34). The system network is reduced by combining parallel subsystems into a single block, whose reliability is given by (34), while series subsystems are reduced to a single block whose reliability is given by (30). This is conceptually analogous to combining series and parallel resistances in electrical circuits. The network is continually reduced until only a single block remains whose reliability is known from all of the subsystem combinations.

To illustrate this, consider the system in Figure 8.13. To determine the reliability, start by combining the three parallel systems S_2, S_3, and S_4, whose reliability is determined by application of (34) to be $R_{s_{2-4}}(t) = 1 - (1 - R_2(t))(1 - R_3(t))(1 - R_4(t))$. The result is then combined with S_1 in series to give the overall system reliability, $R_s(t) = R_1(t)[1 - (1 - R_2(t))(1 - R_3(t))(1 - R_4(t))]$. The chapter concludes with Example 8.8 that addresses combination system reliability.

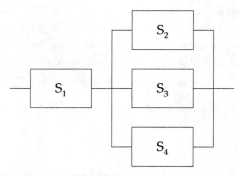

Figure 8.13 A combination series-parallel system. S₂, S₃, and S₄ are redundant parallel systems.

Example 8.8 Combination system reliability.

Problem: Consider the system shown below with the following reliabilities at a fixed time t, $R_1 = R_2 = 80\%$. Determine the reliability that subsystems R_3 and R_4 must have so that the overall system reliability is greater than 95%.

Solution: The parallel systems can combined into single systems whose reliabilities are

$$R_{s_{1,2}} = 1 - (1 - R_1)(1 - R_2)$$
$$R_{s_{3,4}} = 1 - (1 - R_3)(1 - R_4).$$

They are combined in series to give the overall system reliability

$$R_s = \left[1 - (1 - R_1)(1 - R_2)\right] \times \left[1 - (1 - R_3)(1 - R_4)\right].$$

Substituting values and assuming $R_3 = R_4$ gives

$$0.95 = \left[1 - 0.2^2\right] \times \left[1 - (1 - R_{3,4})^2\right].$$

Solving for the reliabilities gives the final result

$$R_3 = R_4 = 0.90.$$

This example demonstrates the power of redundant systems. S_1 and S_2 have somewhat low reliabilities relative to the overall system goal, but the reliability of the parallel combination of S_1 and S_2 is 96%. It requires a reliability for systems 3 and 4 of $R_3 = R_4 = \underline{90\%}$, while the combined reliability of systems 3 and 4 is 99%.

8.4 Summary and Further Reading

This chapter presented the basics of probability theory and methods for estimating the reliability of components and systems. Failure rate is an important quantity that is determined empirically and provides the rate of failure over the lifetime of a component or system. A mathematical definition of reliability was derived from this quantity, which takes a simple exponential form in the case of a constant failure rate. This was applied to estimate the reliability of single components, particularly using failure rates from MIL-HDBK-217F. Issues of thermal transfer and power derating were considered. Reliability estimation was extended to more realistic systems consisting of multiple components in series and parallel forms. The use of redundancy with parallel systems to increase the overall system reliability was addressed.

There are plenty of good textbooks available on the probability theory, if it is necessary to study probability theory further. The book Practical Reliability of Electronic Equipment and Products [Hna03] provides detailed coverage for electrical systems reliability. It includes factors not considered here such as thermal management on printed circuit boards, procurement practices, and electromagnetic interference. Two excellent articles that demonstrate the application of design for reliability and redundancy for an embedded system application are by George Novacek in Circuit Cellar magazine [Nov00, Nov01].

8.5 Problems

8.1   Consider a random variable that obeys a uniform density and varies from 2 to 5. (a) Determine the mean and variance of the random variable. (b) What is the probability that the random variable is between 2 and 3? (c) Plot the CDF.

8.2   In Figure 8.7 it was assumed that the CDF function is monotonically increasing. That is, $F(t) < F(t + \Delta t)$. Show why this is so using equation (13).

8.3   Describe what is meant by *failure rate, failure function,* and *reliability.*

8.4 Consider an integrated circuit that has $\lambda = 50/10^6$ hours . (a) Determine the mean time to failure. (b) Determine the reliability in 5, 10, 15, and 20 years.

8.5 Consider a CD4001BC 2-Input Quad NOR gate (datasheet available in Appendix C). Also assume it is a glass-sealed dual inline package, 15 years in production, $\theta_{JA} = 70°C/W$, the power dissipated in the application averages 10mW, it has B-1 quality, and that it is to be operated in laboratory equipment at an ambient temperature of 25°C. Use the MIL-HDBK-217F data in Appendix B to determine the MTTF and estimate its reliability in 25 years.

8.6 Use the MIL-HDBK-217F data in Appendix B to estimate the reliability of a 32 bit CMOS microprocessor. Assume that it is used in a missile launcher, the ambient temperature is 120°C, that it has 64 pins, that it has been in production for 6 months, B-1 quality parts are used. Determine the MTTF and reliability for the microprocessor in 20 years. (Note: this is the same operating environment used for the BJT in Example 8.5.)

8.7 Consider a 1N4001 diode (datasheet in Appendix C) that is to be operated in the switching circuit shown below. Also assume that the part quality is Lower, and that it is to be used in an airborne inhabited cargo environment at an ambient temperature of 50°C. Use the MIL-HDBK-217F data in Appendix B to determine the MTTF and estimate its reliability in 25 years.

8.8 Use the MIL-HDBK-217F data in Appendix B to determine the reliability of the inverting op amp circuit, shown below, in 15 years. Assume that it is used in an automotive application (environmental factor = G_M), the ambient operating temperature is 80°C, industrial quality parts are employed, and ¼ watt fixed composition resistors of the lowest quality are used. The datasheet for the LM741 op amp is in Appendix C. The LM741 is considered to be a linear microcircuit, comes in a dual inline package (DIP), contains 25 bipolar transistors, has S quality, and has been in production for well over 20 years.

8.9 Consider the circuit in Example 8.4. Assume that a heat sink is attached to the 2N3904 BJT (datasheet in Appendix C) and it has the following thermal resistance values $\theta_{CS} = 10°C/W$ and $\theta_{SA} = 8°C/W$. If the device is operated at an ambient temperature of 130°C, determine the maximum power dissipation of the BJT and its reliability in 50 years.

8.10 Your company intends to design, manufacture, and market a new RAID (Redundant Array of Independent Disks) for network servers. The system must be able to store a total of 500GB of user data and must have a reliability of at least 95% in 10 years. In order to develop the RAID system, 20GB drives will be designed and utilized. To meet the requirement, you have decided to use a bank of 25 disks (25x20GB=500GB) and utilize a system redundancy of 4 (each of the 25 disks has a redundancy of 4). What must the reliability of the 20GB drive be in 10 years in order to meet the overall system reliability requirement?

8.11 S_1 has a failure probability of 2% and S_3 has a failure probability of 3%. $S_{2,1}$, $S_{2,2}$, $S_{2,3}$, and $S_{2,4}$ are identical redundant systems. Determine the required reliability of the redundant systems necessary for the overall system to have a reliability of 94%.

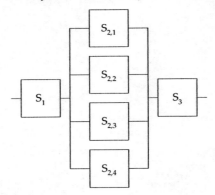

8.12   Consider the design of a triply-redundant majority voting system, with three binary inputs a, b, and c shown below. The inputs represent data from 3 independent sources from which the objective is to determine if the majority of the input bit values are logic level 0 or 1. Each majority circuit outputs the bit value that is in the majority of the inputs. The output of each majority circuit is fed into a resistor and LED (light emitting diode) network. The LEDs are lit if the output of the majority circuit is a logic 1, otherwise the LED is off. Ideally, all three LEDs are lit if the majority of inputs is 1, else they are all off. However, if part of the system fails, the LED readings may not be reliable, so a majority rules decision is used on the LEDs. The criterion used is that if two or more LEDs are on, then the majority of inputs are considered to be high, otherwise, the majority of inputs are considered to be low. Determine the probability of a false reading based upon this criterion if each component in the system (gate, resistor, or LED) has a reliability of 90%.

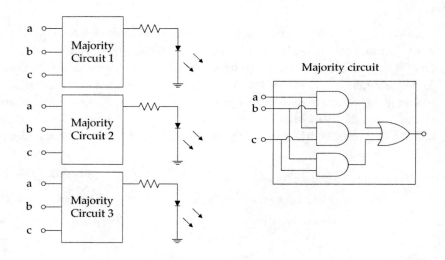

Part III – Professional Skills

Chapter 9 Teams and Teamwork

The way a team plays as a whole determines its success. You may have the greatest bunch of individual stars in the world, but if they don't play together, the club won't be worth a dime. — Babe Ruth

Engineering projects are often very large and require too many diverse skills to be completed by a single person. This necessitates the use of teams, and it is likely that you will participate on many during your professional career. That is why employers of engineering graduates consistently rate teamwork ability as one of the most desirable attributes in a potential employee. Teams can collectively perform tasks that could not possibly be completed by individuals working in isolation. However, just because a team is created does not mean that it will function effectively, and examples of failed teams abound. When people learn that they are going to work on a team, their initial reaction is often one of dread, usually due to bad experiences in the past. The good news is that with some attention to team principles and a conscious effort, working on a team can be a positive experience. Much research has been done and much has been written on teams and teamwork. The objective of this chapter is to provide a primer on what constitutes a team, present models of team development, identify the characteristics of effective teams, and provide guidance for effective teamwork.

Learning Objectives

By the end of this chapter, the reader should:

- Understand the characteristics that define a team and understand why a team is formed.
- Understand different models for the stages of team development.
- Understand the characteristics of effective teams.
- Be able to develop Team Process Guidelines.

9.1 What is a Team?

The concept of a team is used broadly in many different contexts, so the following definition of a team proposed by Katzenbach and Smith in The Wisdom of Teams [Kat93] is provided for the basis of our discussion:

> *A small group of people with complementary skills, who are committed to a common performance, performance goals, and approach for which they hold themselves mutually accountable.*

This concept of a team is highly relevant to engineering design projects.

The definition indicates that a team should be a small group of people, typically two to ten. Teams larger than this become very difficult to manage. The reason is that the number of relationships between people increases rapidly with an increase in the number of people on the team. The number of person-to-person relationships is equal to $n(n-1)/2$, where n is number of people on the team. Any one of these relationships can falter, causing problems. In addition, it is hard to develop consensus on important issues as the number of people on the team increases.

Second, the definition indicates that teams should be composed of members who have complementary skills. Members should be selected based upon the background and skills that each person brings to the team, not based upon personality and friendships. This leads to the related concepts of **cross-functional** and **multi-disciplinary teams.** Cross-functional teams are those that are composed of people from different organizational functions, such as engineering, marketing, and manufacturing [Wil95]. Cross-functional teams are particularly important in new product development, where multiple functions are required to bring a product from concept, to manufacture, and ultimately to the market. Multi-disciplinary implies that the team is composed of members from different disciplines. There is no agreement in the engineering community as to what exactly constitutes a multi-disciplinary team. In general, the idea of complementary skills applies where a team may have representation from multiple technical disciplines. For example, the development of a robot would require members with multi-disciplinary expertise in computing, electronics, and mechanical systems.

Third, the definition states that teams should have common performance goals. The goals take time to develop and the team will likely flounder in creating them, but without specific goals to achieve there is no need for a team. In the context of design, the team's goals are defined by the Problem Statement and Requirements Specification developed in Chapters 2 and 3 respectively. They collectively state what the team is trying to achieve and provide a way for the team to verify its success—are the requirements met and can the team demonstrate that they are?

Finally, the team definition indicates that teams should have a common approach for which they hold themselves mutually accountable. This can be a common approach to solving problems, such as application of the design process. It can also be in terms of a method for holding the team members accountable. One of the biggest problems that teams face is handling situations where the team fails to meet a given objective. This may be due to one or more members failing to meet their commitments. The team must be able hold each other mutually accountable in a constructive manner in these situations.

9.2 Models of Team Development

Bruce Tuckman developed a model for team development that contains fives stages, known as forming, storming, norming, performing, and adjourning [Tuc65]. It is instructive to have an understanding of these stages so that you are aware of the dynamics that occur as a team develops. The five stages are described as follows:

- *Forming.* This is the stage where the team is created. The team members may not know each other, leading them to be anxious and uncomfortable. It is likely that the team's objectives are not well-defined and that the members' roles are ambiguous.

- *Storming.* In this stage, the team works to develop its objectives, while the members try to define their roles. Conflict often appears as members jockey for position and advantage. It is characterized by fighting and power struggles, which may be overt, but in many cases are subtle and below the surface. The team must resolve individual versus group goals. For example, what if the team is trying to select a project concept? Some members may have strong opinions and preconceived notions about what the project should be. Failure to navigate the storming stage means that the team will not reach the performing stage and will be forced to return to the storming stage at some time.

- *Norming.* In this stage the team starts to become more cohesive. Members accept the team's objectives, their roles on the team, and they agree upon procedures. In this phase the team develops common approaches for solving problems and managing conflicts.

- *Performing.* In this stage, the team focuses on performing tasks and achieving the objectives set forth. The team should be collaborating and easily making decisions. Disagreements that arise are accepted and resolved by the team norms.

- *Adjourning.* Here, the team dissolves, hopefully due to successful project completion. However, it can occur abruptly if the project is cancelled, or even worse, if the team is unable to function together.

Katzenbach and Smith proposed the team performance curve in Figure 9.1, which shows the performance impact of teams versus varying levels of effectiveness. The types of teams they defined are the working group, the pseudo-team, the potential team, and the high-performance team. The most closely related stages of team development from the Tuckman model are indicated in parentheses on the figure. Notice that the scale on the x-axis is team effectiveness, not time. There may be an implicit assumption that teams will move through these stages as a function of the time the team is together. That is certainly a desired goal, but not necessarily what the model implies.

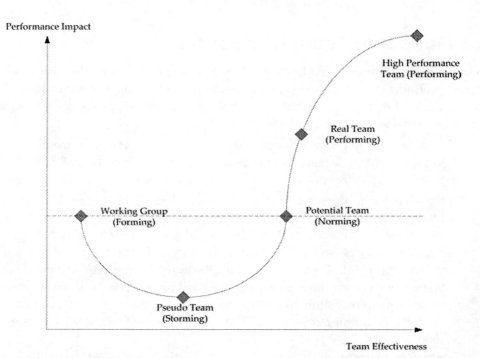

Figure 9.1 The team performance curve proposed by Katzenbach and Smith [Kat93]. This shows the performance impact of teams versus varying levels of effectiveness. The Tuckman stages of team development (forming, storming, norming, performing) that are most closely related to the performance points are superimposed on the model [Tuc65].

The *working group* is defined as a group of individuals working in isolation, who come together occasionally to share information. In other words, if every member of the team works in isolation and meets only to share ideas, they would achieve the performance of the working group. This level of performance is represented by the dashed baseline. In fact, the working group is not a team as given by the definition and serves a very different purpose. The *pseudo-team* represents an under-performing team where the sum effort of the team is below that of the baseline performance, while the *potential team* is one in which the team is functioning at a level equal to that of the working group. At a minimum, teams should at some point perform above the level of a potential team, otherwise, there is no need for a team. The objective is to perform at the level of a *real team,* where the performance exceeds that of the working group. In rare instances the level of a *high-performance team* may be realized in which the team significantly outperforms all similar teams.

9.3 Characteristics of Real Teams

How can a team reach the performing stage and simultaneously become a real team? Unfortunately, there is no set process for guaranteeing that a team will become a real team. There are characteristics and principles of effective teams, and it has been observed that real teams apply these principles. However, realize that there is a difference between a team and teamwork principles. Real teams apply teamwork principles but this does not imply that the converse statement is true—that applying these principles will result in a real team. The lesson is that successful teams adhere to good teamwork principles, but the application of good teamwork principles does not automatically guarantee that teams will be successful. However, ignoring teamwork principles almost always leads to failure. The remainder of this section presents some of the characteristics of real teams.

9.3.1 Select Members Based Upon Skills

Selecting team members based upon their skills is a key to success identified by Katzenbach and Smith. They defined three categories of relevant skills: 1) technical and functional, 2) problem-solving, and 3) interpersonal. The right mix of technical skills on a design project is important to achieve the desired objectives, reinforcing the need for multi-disciplinary and cross-functional teams. Different problem-solving approaches and the mix of interpersonal skills are important as well, but harder to determine. The Meyers-Briggs personality tests or the Keirsey Temperament sorter [Kei84] are tools that can be used to identify different problem-solving approaches and personality traits, but in themselves are not the answer for selecting teams.

One important point to consider is whether teams should be formed by self-selection or assigned by a supervisor. There are pro and con arguments for each method of selection. An argument for self-selection is that members believe the objectives of the team are important, leading to a higher level of commitment. A potential pitfall is that not enough attention is paid to the skills necessary to complete the project. In the second case, teams are assigned by someone who has an understanding of the skills needed for the project and assigns members accordingly. A potential pitfall is that this may create animosity from team members who are dissatisfied with the project from the outset.

9.3.2 Identify Objectives

Teams are created to realize shared goals, and if the goals are not well-defined, the motivation for being part of the team is not clear. In the context of engineering design, the combined Problem Statement and Requirements Specification presented in Chapters 2 and 3 serve that purpose. The Problem Statement describes what the team is trying to achieve, while the Requirements Specification sets verifiable targets that define success. Both items are so important, that there should be a consensus agreement among the team members on their content. The team should be challenged by targets that are aggressive, yet achievable. Identifying measurable objectives was identified by Katzenbach and Smith as one of the most important attributes of successful teams. Further, all team members must be committed to achieving the objectives.

9.3.3 Develop Decision-Making Guidelines

Teams regularly make decisions that affect all of its members and the success of the project. The team must determine how decisions are to be made, and once they are, all members need to accept the outcome. The team also needs to understand the importance of different decisions—not all are equally important and different approaches should be used depending upon the impact of the decision. Models for decision-making are outlined as follows [Joh02]:

1. *Decision by Authority.* The leader makes all decisions, typically without discussion by the rest of the team. This is effective for fast decision-making, but does often not lead to the best decision. It can also produce resentment among the team.

2. *Expert Member.* The most expert member on the subject decides. This is effective in cases where there is a single member who is clearly the expert on the subject.

3. *Average Member Opinion.* The average team member opinion is used. A method needs to be devised to determine what constitutes the average opinion.

4. *Decision by Authority after Discussion.* The leader makes a decision after all team members discuss the issue and provide input.

5. *Minority Control.* A few members act as a subcommittee and solve the problem. This makes sense if the minority group consists of experts on the particular problem.

6. *Majority Control.* A simple majority is used to make decisions.

7. *Consensus.* All team members must agree to and commit to the decision. This generally comes after much discussion and evaluation of the different alternatives. This is the best approach, and the most time-consuming. It is not necessary for all decisions to be made this way, but consensus should be reached for important decisions.

A formal and widely used method for brainstorming and team decision-making is known as the *Nominal Group Technique* (NGT) [Del71]. NGT was covered in Chapter 4 and is repeated here. In NGT, each team member silently generates solutions to a problem and the ideas are then reported out in a round-robin fashion until all ideas are exhausted. Then, each member gets to cast a predetermined number of votes for the ideas presented. The top idea is then selected, or alternatively, the top few ideas are discussed further and voted upon again. The steps of NGT are outlined below:

1. *Read problem statement.* It should be read out loud by a team member (the facilitator).

2. *Restate the problem.* Each person restates the problem in their own words to ensure that all members understand it.

3. *Silently generate ideas.* All members silently generate ideas during a set period of time, typically 5–15 minutes.

4. *Collect ideas in a round-robin fashion.* Each person presents one idea in turn until all ideas are exhausted. The facilitator should clarify ideas and all should be written where the entire team can view them.

5. *Summarize and rephrase ideas.* Once the ideas are collected, the facilitator leads a discussion to clarify and rephrase the ideas. This ensures that the entire group is familiar with them. Related ideas can be grouped or merged together.

6. *Vote.* Each person casts a predetermined number of votes, typically three to six, for the ideas presented. The outcome is a set of prioritized ideas that the team can further discuss and pursue.

9.3.4 Hold Effective Meetings

Teams need to meet for a variety of reasons, such as determining objectives, tracking progress, assigning tasks, preparing deliverables, and resolving problems. It is in these meetings that teamwork principles are most critical and where problems can arise. In the workplace it is estimated that people spend half of their time in tasks that are related to meetings (preparing, attending, and following up). Three elements of effective meetings are: 1) well-defined roles for the participants, 2) a structure for conducting the meeting, and 3) the application of interpersonal skills [Bel94]. Poorly organized and ineffective meetings lead to cynicism and poor team performance. The structure of meetings does not always need to be the same. Some keys to effective meetings are:

- *Have an agenda.* Agree upon the goals for a meeting in advance.

- *Show up prepared.* All members should show up on time with their tasks completed.

- *Pay attention.* Each person should speak in turn, and there should be no disruptive side conversations. No person should monopolize the conversation and all points of view should be heard.

- *Agree upon a meeting time and place.* People have busy schedules and different work habits. Believe it or not, failure to agree upon this basic issue can lead to tremendous conflict.

- *Summarize.* At the end of the meeting summarize what was discussed, important decisions made, and actions to be taken.

9.3.5 Develop Team Roles

As the group finds its norms, members should settle into different roles. They may be based upon the technical aspects of the design, such as hardware, software, and mechanical systems. The team may find that certain members are more effective at different tasks such as procuring parts, making presentations, and writing technical documents. Roles often evolve and change as the team develops.

Team roles are usually not known early on in the forming stage and members can be assigned formal roles to perform in at the outset. This is also common in group problem solving sessions and a typical set of roles follows [App01]:

- *Leader.* The leader is not necessarily the authority figure, but should guide the team through the processes of decision-making and problem-solving. Effective leaders should ensure that all opinions are heard and that everyone is involved. Leaders coordinate tasks such as identifying the agenda.

- *Recorder.* The recorder is responsible for maintaining a record of the team's work, documenting important results, and recording responsibilities for different tasks.

- *Spokesperson.* The spokesperson articulates the team's results. The recorder and spokesperson roles are sometimes combined.

- *Optimist.* The optimist's role is to examine reasons why ideas and concepts will work, find their merit, and advocate other's ideas.

- *Pessimist.* The pessimist should challenge ideas and assumptions, and make sure that opposing ideas are presented and discussed.

- *Analyst.* The team analyst performs the important role of observing the team processes and provides feedback to the team for improvement.

The roles presented here are by no means definitive and other models having different roles can be utilized. The common thread is to have participants gain experience in the different roles so that they can become more effective team members. No matter what role a member is playing, they should all contribute to the solution of the problem at hand. As the team evolves and becomes more experienced, the need for formal roles will diminish.

9.3.6 Assign Tasks and Responsibilities

Nothing creates problems more than when team members do not have clearly defined responsibilities and tasks. The workload needs to be distributed in a fair manner and all members must perform real work. Without this, one or two people do most of the work, or even worse, nothing gets done. Chapter 10 presents project management principles, one of the primary aims of which is to develop and assign tasks to team members. Teams may consider designating one person as the project manager who is responsible for tracking the team's progress and ensuring that members are meeting their commitments. It is important that the project manager also be assigned tasks, outside of project management, that contribute to the project.

9.3.7 Spend a Lot of Time Together

Teams that spend a lot of time together are generally more successful [Kat93]. This is not only time together in meetings and working on project deliverables, but also includes extracurricular activities. All members of the team should be included, not just a subgroup. Experience shows that on teams of three students, two members may spend a great deal of time together and form a bond. This can alienate the third person (not necessarily intentional) and lead to conflict.

9.3.8 Respect Each Other

Imagine a scenario where you are assigned to a team and one of the members is that quiet, weird guy who never says anything, or that loudmouth who never shuts up. You may have a real challenge ahead of you, but working with people who are not our best friends is a reality of life. You have to learn how to deal with personality issues if you want to be effective on teams. One key is to demonstrate respect for the other members of your team. Ways to do this are to:

- *Listen actively.* Most people are not effective listeners. As they listen to others they are mentally formulating their response, while not actually listening to the other person. By listening to the other person, then formulating an appropriate response, you will develop a better understanding of their viewpoint and demonstrate that their opinion is being heard.

- *Consider how you respond to others.* How you respond to others impacts your effectiveness. Are you negatively evaluating others ideas or treating their ideas unfairly?

- *Constructively criticize ideas, not people.* It is fine to examine problems and constructively criticize ideas. Consider looking for ways to improve upon ideas that have flaws.

- *Respect those not present.* Nothing creates divisions and factions in a team faster than personally criticizing members not present. If a team member is not performing, refer to the Team Process Guidelines examined in Section 9.4 for holding them accountable.

- *Communicate your ideas.* If you have an idea, clearly state it and be prepared to explain what the merits are. Be prepared for critical analysis and discussion of the idea.

9.3.9 Manage Conflicts Constructively

Conflicts inevitably occur on teams, and the more constructively they are addressed the better. Real teams do encounter conflicts, but they are adept at resolving them. Conflicts can be good, particularly when they lead the team to the solution of a problem or the development of consensus. They also provide an opportunity for members to express their opinions on important issues. Left unresolved and unspoken, they lead to resentment, suspiciousness, and issues. Left unresolved and unspoken, they lead to resentment, suspiciousness, and escalate into personal conflicts. Conflicts may be false (you misinterpret another's behavior), based on performance (member(s) aren't doing their work, or it is of poor quality), or based upon procedure (disagree with the way that meetings are run or decisions are made) [Dom01]. Some strategies for resolving conflicts are:

- *Focus on performance and ideas.* Focus on the performance of the team and not individual personality. Also, focus on ideas and constructively criticize them.

- *Listen to others.* It is very important to remain calm and listen carefully to others when conflicts arise.

- *Identify concerns.* If you have concerns about something, it is best to identify and address them, rather than hiding them.

- *Apply the team's process guidelines.* In the next section, a format for process guidelines that govern the behavior and processes of a team is given. This should address a team's approach for resolving conflicts. Importantly, the team must remember to apply the guidelines when conflicts arise.

- *Develop a plan to resolve the conflict.* Again, conflict can be positive and lead to the solution of problems. The team may not be able to resolve a conflict immediately, but can develop a plan for solving it within a specified period of time.

- *Mediation.* A mediator can be used after all avenues for resolution of the conflict have been exhausted. One way to do this is to employ a variation of the Delphi Technique. The Delphi Technique was originally developed as a brainstorming method to generate ideas anonymously from experts outside of a team or organization. It is applied for mediation through the following steps:

  1. Each member of the team should anonymously supply a description of the conflict and suggested remedies to the mediator.

  2. The mediator proposes a solution to the conflict, which is fed back to the team.

  3. The team members are given an opportunity to suggest modifications to the proposed resolution.

  4. Steps 1-3 are repeated until consensus is achieved.

9.4 Project Application: Team Process Guidelines

Teams should develop clear and detailed guidelines that govern their processes. These guidelines are based on the items covered in the previous section. The entire collection of guidelines forms the *Team Process Guidelines*. Issues addressed in the Team Process Guidelines include:

- *The team's name.* This creates an early opportunity for decision-making.

- *The team's mission and objectives.* This can be a brief re-statement of the design problem statement.

- *Decision-making guidelines.* Indicate what techniques are going to be used to make decisions and when they will be applied.

- *Meeting guidelines.* How will meetings be run and what are the expectations for team members in the meetings.

- *Team roles.* If teams self-select, they should justify their team choice with identification of the complementary skills each member brings to the team.

- *Conflict resolution.* Identify how the team will resolve conflicts. For example, what will the team do when members are doing sub-standard work and not meeting the performance objectives? How will the team determine what sub-standard performance is? What happens if a member misses a meeting? What will the team do if they cannot resolve the conflict among themselves?

This document should be developed early on, referred to regularly throughout the project, and updated by the team as necessary. Table 9.1 contains a checklist and self-assessment of the team formation stage and processes.

9.5 Summary and Further Reading

This chapter touched on some of the fundamental concepts of team development that have been reported in the literature. Two well-known models of team development were presented. The first is the Tuckman model of the forming, storming, norming, performing, and adjourning stages. It is important for teams to carefully navigate the formative stages, agree upon the objectives, and reach the performing stage. The second is the Katzenbach and Smith team performance curve presented in Figure 9.1. Selected characteristics of real teams were identified. Finally, Team Process Guidelines are a tool for identifying a team's norms, problem solving approach, and self governing principles. A format for the guidelines was presented.

There are many excellent books and resources available that examine teams and teamwork. One that influenced this chapter greatly is <u>The Wisdom of Teams</u> by Katzenbach and Smith [Kat93]. It is a seminal work in the field and a widely recognized resource that addresses how to create effective teams. The forming, storming, norming, performing, and adjourning model of team development was proposed by Bruce Tuckman [Tuc65] and is widely documented as a good model for team development. <u>The Team Training Workbook</u> [Bel94] is another valuable resource developed by Arizona State University engineering faculty members. <u>Teamwork and Project Management</u> by Karl Smith [Smi04] is a short introduction to both teamwork principles and project management for engineering students. Free personality tests and temperament sorters can be found online; one example being the Keirsey Temperament Sorter at <u>www.advisorteam.com</u>.

Table 9.1 Checklist and self-assessment for team formation and processes. 1 = Strongly Disagree, 2 = Disagree, 3 = Neutral, 4= Agree, 5 = Strongly Agree.

Team Formation	Score
The team's objectives are clearly defined.	
There is consensus among all team members that the objectives are the correct ones.	
The team members complementary skills (technical, functional, interpersonal) have been identified.	
There are enough members on the team to cover all of the necessary competencies.	
There are not too many members on the team.	
Team Processes	
The team has developed clear guidelines for resolving conflicts and disagreements.	
The team has developed effective guidelines for holding all members of the team mutually accountable for achieving the objectives.	
The team has developed a strategy for holding effective meetings.	
The team has agreed upon a mutual meeting time and place.	
The team members trust each other.	
The team members demonstrate respect for each others ideas.	

9.6 Problems

9.1   Explain the difference between cross-functional and multi-disciplinary teams.

9.2   Identify the characteristics of the forming, storming, norming and performing stages in team development.

9.3   Describe the distinction between teams and teamwork.

9.4   According to this chapter, it is difficult to develop a consensus as the number of team members increases. Consider the situation where a team needs to agree on a proposal. Furthermore, assume that each team member's vote is random with a 50% chance of agreeing with the proposal. Plot the probability of the team unanimously agreeing to the proposal versus the number of team members. Consider team sizes from 2 to 10. Overlay three additional plots for the situation where each team member has a 75%, 90%, and 99% chance of agreeing to the proposal.

9.5   **Project Application.** Develop Team Process Guidelines as proposed in Section 9.4.

9.6   **Project Application.** Complete the team self-assessment in Table 9.1.

Chapter 10 Project Management

If you fail to plan, then you plan to fail.—Anonymous

The engineering community has led in the development of project management practices because building complex systems is a tremendous technical and managerial challenge. Currently, businesses tend to organize around projects that have significant value to the organization. Consequently, project management is consistently rated by employers as one of the most desirable skills sought in new college engineering hires [Par03]. The project management field includes topics such as initiating a project, team management, cost management, risk management, controlling, resource management, and performance management, to name a few. Many of these are addressed throughout this book from an engineering design viewpoint such as controlling (design process), initiating (project selection), performance management (requirements and testing), and team management.

The three important objectives of project management are to complete projects that are on-time, within budget, and meet the requirements of the user. Since user requirements were addressed in Chapters 2 and 3, this chapter addresses the remaining two objectives of time and cost management. Time management introduces the work breakdown structure, which identifies the activities (combined tasks and deliverables) required to complete the project. Responsibility for completing the activities is then assigned to members of the team. Two graphical representations of the work breakdown structure, the network diagram and the Gantt chart, are introduced. These visual depictions show dependencies between tasks and allow for a quantitative analysis of the project plan. The chapter concludes with methods of cost estimation.

Learning Objectives

By the end of this chapter, the reader should:

- Be able to create a work breakdown structure.
- Be able to create network diagrams and Gantt charts.
- Be able to determine the critical path for completing a project and the float time for each activity in the plan.
- Be able to conduct break-even analysis and understand some basic methods of cost estimation.

10.1 The Work Breakdown Structure

The *work breakdown structure* (WBS) is a hierarchical breakdown of the tasks and deliverables that need to be completed in order to accomplish the project objectives. Creating the WBS is typically the first step in project planning. A WBS is an ordered set of activities required to complete the project. An *activity* is a combination of a task and its associated deliverables. *Tasks* are actions that accomplish a job, while *deliverables* are entities that are delivered to the project based upon completion of tasks. Examples of deliverables include a circuit design, a software module, the integration and test of modules, a report, a presentation, or obtaining an approval. Example tasks include conducting research or writing a program.

The concept of the WBS was formalized by the United States Military in the 1993 document <u>Work Breakdown Structure Handbook</u> [MIL-HDBK 881]. The WBS has gained wide acceptance in industry and is described in MIL-HDBK 881 as follows:

- *A product-oriented family tree composed of hardware, software, services, data, and facilities. The family tree results from systems engineering efforts.*

- *A WBS displays and defines the product, or products, to be developed and/or produced. It relates the elements of work to be accomplished to each other and to the end product.*

- *A WBS can be expressed down to any level of interest. However the top three levels are as far as any program or contract need go unless the items identified are high cost or high risk. Then, and only then, is it important to take the work breakdown structure to a lower level of definition.*

This description indicates that WBS results from systems engineering efforts and that the structure of the WBS follows the design hierarchy. The second bulleted item focuses on the activities for the project. Gray and Larson [Gra02] recommend identifying the following activity attributes for each activity:

- A definition of the work to be done or delivered.
- A timeframe for completion of the activity.
- Resources needed to complete the activity.
- Person(s) responsible for the activity.
- Predecessors (or dependencies) for the activity. Predecessors are other activities that must be completed before the work can start.
- Checkpoints for monitoring progress.

The collection of activities and their attributes are gathered in the WBS table. The rows of the WBS table represent the project activities. These activities are arranged in a hierarchical fashion; each major activity is followed by its constituent sub-activities. The columns of the WBS table are the activity attributes proposed by Gray and Larson. Table 10.1 contains the WBS table for the temperature monitoring design examined in Chapter 5 (Section 5.7).

Table 10.1 Example work breakdown structure for the design of a temperature monitoring system.

ID	Activity	Description	Deliverables / Checkpoints	Duration (days)	People	Resources	Prede-cessors
1	**Interface Circuitry**						
1.1	Design Circuitry	Complete the detailed design and verify it in simulation.	• Circuit schematic • Simulation verification	14	Rob (1) Jana (1)	• PC • SPICE simulator	
1.2	Purchase Components		• Identify parts • Place order • Receive parts	10	Rob		1.1
1.3	**Construct & Test Circuits**	Build and test.					
1.3.1	Current Driver Circuitry	Test of circuit with sensing device.	• Test data • Measurement of linearity	2	Jana (1) Rob (2)	• Test bench • Thermo-meter	1.2
1.3.2	Level Offset & Gain Circuitry	Test of circuit with voltage inputs.	• Test data • Measurement of linearity	3	Rob (1) Jana (2)	• Test bench	1.2
1.3.3	Integrate Components	Integrate the current driver and offset circuits.	• Test data verifying functionality and linearity requirement	5	Rob (1) Jana (1)	• Test bench • Thermo-meter	1.3.1 1.3.2
2	**LED & Driver Circuitry**						
2.1	Research A/D Converters	Make selection of A/D converter.	• Identify types, cost, and performance • Identify two potential converters for purchase	1	Alex	• Internet	
2.2	Complete Hardware Design	Design conversion hardware.	•Circuit schematic • Simulation verification	7	Ryan (1) Alex (2)	• Digital circuit simulator	2.1
2.3	Purchase LED & Driver Components		• Identify parts • Place order • Receive parts	10	Rob		2.2
2.4	Construct & Test	Test with supply voltage input.	• Test data showing digital output vs. voltage inputs	5	Alex (1) Ryan (2)	• Test bench • Logic analyzer	2.3
3	System Integration & Test	Complete integration of front-end and LED driver circuitry.	• Test data demonstrating functionality from temp input to LED output • System linearity measurement	7	Alex (1) Rob (1) Jana (1) Ryan (1)	• Test bench • Digital logic analyzer • Thermo-meter	1.3.3 2.4 (or 1 and 2)

Each activity is assigned an identification number (ID) and a name. The hierarchical nature of the activities is reflected in the ID numbering scheme. The three highest-level activities in the project plan are the Interface Circuitry (1), LED and Driver Circuitry (2), and System Integration and Test (3). The first two activities are further refined into sub-activities shown by the indented items. For example, the Interface Circuitry activity contains three sub-activities: Circuit Design (1.1), Purchase Components (1.2), and Construction and Test (1.3). The Construction and Test activity is further subdivided, producing a total of three levels in the hierarchy. In this example, the WBS follows the hierarchical breakdown of the design architecture itself.

Descriptions and deliverables for each activity are provided. The deliverables also serve as checkpoints for monitoring the activity. The identification of checkpoints becomes more important as the complexity and duration of an activity increase. The deliverable items identified for this example include circuit design schematics, simulation results, and test data. Deliverables are even defined for activities that can be somewhat nebulous, such as Research A/D Converters (2.1). Doing so ensures that activities do not become open-ended and have specific deliverables for monitoring their progress.

The fifth column in Table 10.1 is the estimated duration for each activity. The ability to estimate durations is heavily influenced by past experience. The general tendency is to underestimate the amount of time required. Keep in mind that credibility is lost if delivery of the final system is repeatedly delayed, so durations should be estimated as accurately as possible. Take into account time for unexpected problems, such as equipment failure, delayed delivery of parts and equipment, and illness. System integration and testing are tasks that are notoriously time-consuming. A method for estimating activity duration comes from the Project Evaluation Review Technique (PERT) developed by the US Navy in the 1950s. Empirical studies show that durations typically follow a Beta probability density. Based upon this model, the duration of an activity is estimated as

$$t_e = \frac{t_a + 4t_m + t_b}{6}, \tag{1}$$

where t_a is the most optimistic time estimate, t_b is the most pessimistic time estimate, and t_m is the most realistic time estimate. The advantage of this is that it forces one to consider best and worst-case scenarios in the plan.

The person(s) responsible for each deliverable are identified in the WBS. Teams need to consider how to assign responsibility to ensure mutual accountability is achieved. Assigning one person to an activity provides a single person who is responsible for it. However, what if that person needs additional support or backup? Who is going to help them if they are unable to deliver? What if a team member becomes ill? It may make sense to assign multiple people to activities based upon complexity. In Table 10.1, when a number is placed adjacent to a person's name, it indicates that the person has primary (1) or secondary (2) responsibility for that activity.

The resources needed to complete the activity include material, equipment, and labor. Material includes items that are consumed in creating the deliverable, such as electronic components and printed circuit boards. Examples of equipment include test equipment, computers, and software.

Predecessors for an activity are the activities that must be completed before the given activity can begin. Identifying predecessors is necessary for determining the sequencing of the activities and the time to complete the project.

10.2 Network Diagrams

A *network diagram* is a directed graph representation of the activities and dependencies between them for a project. Activity B is dependent on activity A when A must be completed before B can be started. The network diagram allows a graphic visualization of the project that also allows for quantitative analysis. In the *Activity on Node* (AON) form, the activities are represented by nodes and the dependencies by arrows. Examples of the basic connections allowed in the AON representation are shown in Figure 10.1.

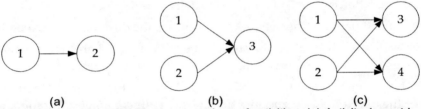

(a)   (b)   (c)

Figure 10.1 Activity on Node (AON) representations of activities. (a) Activity 1 must be completed before activity 2 can begin. (b) Activities 1 and 2 must be completed before activity 3 can begin. (c) Activity 1 and 2 must be completed before activities 3 and 4 can begin.

An example network diagram for a simple project is shown in Figure 10.2. Dummy activity nodes for the start and end of the project have been included. The ID number and estimated duration in days are indicated inside each node. A given network diagram will have multiple paths from the start to the end of the project. A path is any connected sequence of activities from the start node to the end node. In this example, there are four paths to completion: $P_1 = \{1,4\}$, $P_2 = \{1,5\}$, $P_3 = \{2,3,4\}$, and $P_4 = \{2,3,5\}$. The completion of an individual path does not result in completion of the project—all paths (and consequently all activities) must be completed for the entire project to finish.

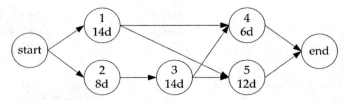

Figure 10.2 Example network diagram. Each activity contains the activity ID and the estimated duration in days.

The duration of each path is determined by summing the duration of all activities on the path, which in this case is 20, 26, 28, and 34 days for paths P_1 to P_4 respectively. The path with the longest duration is known as the *critical path* since it represents the minimum time required to complete the project. In this example the critical path is P_4. Activities on the critical path are of particular interest, since, if they fall behind schedule, or experience *slippage*, the overall completion time of the project is delayed. The other paths can become critical paths if their activities experience sufficient slippage to form a new critical path. A quantity known as *float* quantifies this margin. Float is the amount of time an activity can slip without extending the overall completion time of the project. Thus, all activities on the critical path have zero float by definition. The following method is utilized to determine the float for activities:

1. Identify all paths on the network diagram and the duration for each path.

2. The path with the longest duration is the critical path. Label the critical path duration as t_{cp}. All activities on the critical path have zero float.

3. To determine the float for an activity that is not on the critical path, find all paths that the activity lies on and identify the one with the longest duration. Label this as t_{lp}. The float for the activity is calculated as

$$float = t_{cp} - t_{lp}.$$

(2)

Examples 10.1 and 10.2 examine float time computation.

Example 10.1 Float time calculation.

Problem: Calculate the float for activities in the network diagram shown in Figure 10.2.

Solution: The first two steps of the process have already been completed, which are the identification of the paths, their duration, and the critical path. To summarize, the paths are $P_1 = \{1,4\}$, $P_2 = \{1,5\}$, $P_3 = \{2,3,4\}$, and $P_4 = \{2,3,5\}$ with durations of 20, 26, 28, and 34 days. The critical path is P_4 and $t_{cp} = 34$ days, so activities 2, 3, and 5 have zero float. Thus

$$float_{2,3,5} = \underline{0 \text{ days}}.$$

The only two activities that are not on the critical path are 1 and 4. Let's examine activity 1 first. It lies on paths P_1 and P_2 with durations 20 and 26 days, thus the longest path to completion, t_{lp1}, for activity 1 is 26 days. The float is calculated from (2) as

$$float_1 = t_{cp} - t_{lp1} = 34\text{-}26 \text{ days} = \underline{8 \text{ days}}.$$

Activity 4 lies on P_1 and P_3, so $t_{lp4} = 28$ days and the float is

$$float_4 = t_{cp} - t_{lp4} = 34\text{-}28 \text{ days} = \underline{6 \text{ days}}.$$

This means that activity 1 can slip by 8 days and activity 4 can slip by 6 days without impacting the time to complete the project.

Example 10.2 Network diagram construction and float time calculation for the temperature display.

Problem: For the example WBS in Table 10.1, (a) create a network diagram, (b) determine the critical path and project completion time, and (c) determine the float for all activities not on the critical path.

Solution:
(a) The network diagram is constructed from the dependencies identified in Table 10.1 and is shown below.

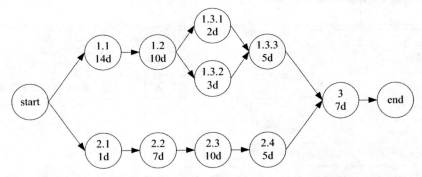

(b) The three paths from start to end are: $P_1 = \{1.1, 1.2, 1.3.1, 1.3.3, 3\}$, $P_2 = \{1.1, 1.2, 1.3.2, 1.3.3, 3\}$, and $P_3 = \{2.1, 2.2, 2.3, 2.4, 3\}$ which have durations of 38, 39, and 30 days respectively. Thus the <u>critical path is P_2</u> and its duration is

$$t_{cp} = \underline{39 \text{ days}}.$$

(c) All activities on path P_1 are also part of the critical path, with the exception of activity 1.3.2, which has

$$float_{1.3.2} = 39\text{-}38 \text{ days} = \underline{1 \text{ day}}.$$

Activities 2.1-2.4 have the collective float

$$float_{2.1-2.4} = 39\text{-}30 \text{ days} = \underline{9 \text{ days}}.$$

The strength of the network diagram is that it provides an intuitive graphical representation of activities and their dependencies on one another. This is particularly valuable for complex projects where the paths to completion may not be obvious. It also allows identification of the critical path and float times for activities. A disadvantage of the network diagram is that it may be difficult to encapsulate the amount of information required for an in-depth project on a single page in an easy to read format.

10.3 Gantt Charts

Gantt charts, developed by a mechanical engineer named Henry Gantt (1861-1919), are a bar graph representation of activities on a timeline. An example Gantt chart is shown in Figure 10.3 for the temperature display design. The Gantt chart effectively shows the WBS and the timeline for completion. A traditional weakness of the Gantt chart has been the inability to show the dependencies between activities. However, as seen in this example, this has been remedied by modern project management software where the dependencies are indicated by the connecting arrows between tasks.

Task Name	Start	Finish	Duration
1: Interface Circuitry	1/10/2005	2/22/2005	32d
1.1: Design Circuitry	1/10/2005	1/27/2005	14d
1.2: Purchase Components	1/28/2005	2/10/2005	10d
1.3: Construct & Test Circuits	2/11/2005	2/22/2005	8d
1.3.1: Current Driver Circuitry	2/11/2005	2/14/2005	2d
1.3.2: Level Offset & Gain Circuitry	2/11/2005	2/15/2005	3d
1.3.3: Integrate Components	2/16/2005	2/22/2005	5d
2: LED & Driver Circuitry	1/10/2005	2/9/2005	23d
2.1 Research A/D Converters	1/10/2005	1/10/2005	1d
2.2 Complete Hardware Design	1/11/2005	1/19/2005	7d
2.3 Purchase LED & Driver Components	1/20/2005	2/2/2005	10d
2.4: Construct & Test	2/3/2005	2/9/2005	5d
3: System Integration & Test	2/23/2005	3/3/2005	7d

Figure 10.3 Gantt chart for the temperature display project created using Microsoft Visio™.

10.4 Cost Estimation

The second main objective of this chapter is to address how to complete projects within budget. In order to do this, the costs associated with the design, development, and manufacture of the system need to be estimated. This section describes break-even cost analysis and economic considerations followed by methods of cost estimation.

10.4.1 Break-Even Analysis

A break-even analysis aims to determine the number of units that must be sold for costs and revenues to be equal—in other words, for there to be no profit or loss. The two types of costs that factor into this analysis are fixed and variable costs. *Fixed costs* are those that are constant regardless of the number of units produced and cannot be directly charged to a process or activity. Examples are rent, overhead, insurance, property taxes, design and development costs, capital expenditures, market research, and sometimes labor costs depending upon the situation. Capital expenditures are costs incurred for long-term assets such as equipment or buildings. *Variable costs* vary depending upon the process or items being produced, and fluctuate directly with the number of units produced. Examples are raw materials, inventory, energy costs, and labor costs.

The *break-even point* is the point where the number of units sold is such that there is no profit or loss. It is determined from the total costs and revenue. The total cost required to produce a product is the sum of fixed and variable costs. Assuming n units are sold, the total cost is

$$\text{total cost} = \text{fixed cost} + n \times \frac{\text{variable cost}}{\text{unit}}. \tag{3}$$

The total revenue generated by the sale of the n units is directly related to the sale price

$$\text{revenue} = n \times \frac{\text{sale price}}{\text{unit}}. \tag{4}$$

The break-even point is where the revenue and total costs are equal

$$\boxed{n \times \frac{\text{sale price}}{\text{unit}} = \text{fixed cost} + n \times \frac{\text{variable cost}}{\text{unit}}}. \tag{5}$$

The break-even analysis is shown graphically in Figure 10.4 with the costs and revenue plotted as a function of units sold. Equation (5) allows different scenarios to be examined. For example, based upon a certain cost structure and sales price, the volume of sales necessary to break-even can be computed. Or, based upon a cost structure and projected number of units sold, a target sales price can be selected. Example 10.3 demonstrates the application of break-even analysis for the development of the Hewlett-Packard (HP) DeskJet printer.

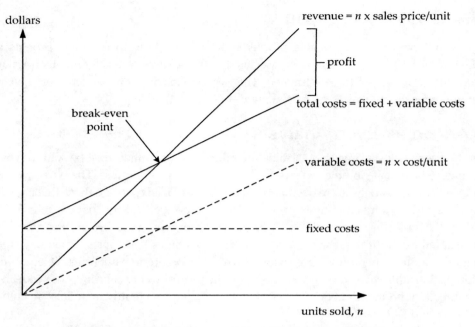

Figure 10.4 Graphical representation of the break-even analysis.

Example 10.3 Break-even analysis for the HP DeskJet printer.

**Problem:** The following data has been publicly reported for the development and sale of the HP DeskJet [Ulr03]: sales price = $300, development cost = $50 million, production investment = $25 million, annual production (sales) volume = 4 million units per year, and the sales lifetime is 2 years. Assuming a fictitious variable production cost of $225/unit, determine: (a) the number of units that must be sold to break even, and (b) the profit expected over an estimated sales lifetime of 2 years.

**Solution:**

(a) The objective is to determine the sales volume, n, necessary to break even. The fixed costs are the sum of the development costs and production investment. So

$$\text{fixed costs} = \$(50+25) \text{ million} = \$75 \text{ million.}$$

This leads to a total cost for n units sold of

$$\text{total cost} = \$75 \text{ million} + n \times \$225 \,.$$

The revenue is

$$\text{revenue} = n \times \$300 \,.$$

Setting the revenue and total cost each equal at the break-even point produces

$$n \times \$300 = \$75 \text{million} + n \times \$225 \,.$$

Solving for the final number of units gives

$$n = \underline{1 \text{ million units}}.$$

(b) Profit is the differential between the total revenue and the total cost and is expressed as

$$\text{profit} = \text{total revenue} - \text{total costs}.$$

For an expected volume of 4 million units per year over 2 years

$$\text{profit} = n \times \$300 - (\$75 \text{ million} + n \times \$225)$$
$$= 8 \text{ million} \times \$300 - (\$75 \text{ million} + 8 \text{ million} \times \$225)$$

$$\text{profit} = \underline{\$525 \text{ million}}.$$

10.4.2 Cost Models

The costs must be accurately estimated in order to realize the expected profit. The goal here is not to address the complete subject of cost estimation, which is beyond the scope of this book, but instead present some basic concepts and techniques for estimating development costs. Many projects go over budget during development, and this is a particularly important consideration from a design viewpoint. The WBS is a valuable tool for cost estimation because it divides the project into manageable pieces whose individual costs can be more readily estimated.

As identified in the WBS, the development costs associated with a project typically include labor, equipment, and materials. Equipment costs can be determined in a fairly straightforward manner, because many of the equipment needs are known *a priori*. Labor costs are tied directly to the length of the project and are often the largest expense.

Estimates of labor costs are usually based upon past experience and expert opinion. This means asking others to estimate the cost and use it as a guide. The estimation formula in (1) for activity duration can be applied for costs as

$$\text{cost} = \frac{\text{cost}_a + 4\text{cost}_m + \text{cost}_b}{6}, \tag{6}$$

where cost_a is the most optimistic cost estimate, cost_m is the most likely cost estimate, and cost_b is the most pessimistic cost estimate.

A more formal approach for estimating labor costs is to use empirical models that represent a quantification of past experience. The models estimate an output based upon quantifiable inputs related to the design or technology. Example inputs are the number of subsystems, the estimated complexity of a circuit design, or an estimate of the lines of code necessary for a software project. This changes the problem from opinion-based estimation to estimation of a quantity that is presumably easier to find and a better indicator of the cost. The output of the model is the cost or another quantity that is directly related to it, such as the estimated number of person-hours.

The simplest example is a linear model ($y=mx+b$) for estimating an output y (cost or person-hours) based upon an input x. For example, IBM modeled software development project costs using the number of lines of code as the input [Jal97] as

$$effort = a \times KLOC + b. \tag{7}$$

The output is an estimate of the effort in person-months and the input is the projected number of lines of code, KLOC. KLOC, pronounced "kayloc", equals thousands (kilo) of lines of code. The linear model was found to work well for relatively small development projects with between 4 and 10 KLOC. As the complexity increased, an exponential model was found to be more realistic where

$$effort = a(KLOC)^b. \tag{8}$$

By observation of 60 software development projects at IBM, the exponential model was fit to observed data and the parameters were estimated to be $a=5.2$ and $b=0.91$. The KLOC model is applied in Example 10.4.

Example 10.4 Effort estimation using KLOC.

Problem: Consider a software development project that has a team of 10 software development engineers. The team has proposed a design and estimates that it will require 50,000 lines of code to complete the project. The average cost to the company for an engineer is $100,000 per year, including salary, benefits, and overhead. Estimate (a) the time required to the complete the project and (b) the labor costs.

Solution:
(a) Based upon the projected value KLOC, the exponential model in (8) is most appropriate

$$effort = 5.2(50)^{0.91} = 183 \text{ person - months}.$$

Since there are 10 developers on the project, the estimated time is determined by dividing the effort by 10, to produce an estimated time of 18.3 months.

(b) The labor costs for development are determined from the number of person-months and the average monthly salary of a development engineer as

$$labor \ cost = 183 \text{ person - months} \times \frac{\$100,000}{year} \times \frac{1 \ year}{12 \ months}$$

$$labor \ cost = \$1.53 \text{ million}.$$

The models in (7) and (8) are simplistic in that there is a single input to the estimator. For example, what if there were 100 engineers assigned to the project in Example 10.4? The model indicates that the project would be completed in 1.83 months at the same cost. Common sense dictates that this is unrealistic; this problem is commonly referred to as the "mythical man-month" (person-month). The mythical person-month refers to the fact that just adding more people to the team will not linearly reduce development time. The reality is that many factors impact the costs, which leads to effort estimation based on many inputs. An example of this is the Constructive Cost Model (COCOMO) that is also used in software development. There are different levels of COCOMO that allow increasingly complex model inputs, such as the type of technology employed, the maturity of the technology, the size of the team, the experience of the engineers, and the timeframe required to complete the project. Not only do the models estimate costs, but they can also be used to estimate how many engineers are needed to complete projects of a given complexity. Although the examples cited here are from the software field, the concepts are general and applicable to the development of many technologies.

Empirical cost models can also be developed for the materials necessary for the manufacture of items. Material estimates are often based upon the expected size and types of technologies used in the manufactured part. For example, a cost model for the manufacture of a printed circuit board may have as inputs the size of the board, the number of layers, and the type of technology used (such as through-hole vs. surface mount).

10.5 The Project Manager

Many engineering teams have a project manager responsible for planning and organizing the project. The project manager may take primary responsibility for developing the WBS, the network diagram and Gantt chart, the cost estimates, and the budget. Although the project manager may have primary responsibility for the project plan, all team members should have input and contribute to development of the plan. The project manager should monitor the checkpoints and deliverables against the plan and develop strategies for reacting to slippage in any activities. The plan should be updated as necessary and the changes communicated to all involved. The project manager may also have primary responsibility for the purchasing of materials and controlling spending. Project managers are not necessarily the boss in the traditional sense and should be viewed as member of the team. As such, it is important for the project manager to also be responsible for completing project deliverables in addition to the project management tasks.

10.6 Guidance

The following is guidance to consider when creating the project management plan, and is particularly relevant for those working on capstone design projects:

- *Build the plan after the design architecture is complete.* The project plan can be created at any point in the design process. Our experience shows that a good time to develop it is after the system design architecture is complete. The design serves as a good guide for developing the WBS.

- *Take the initial time estimates for activities and double them!* Most people tend to significantly underestimate the amount of time it takes to complete an activity. That is because people often have a conceptual idea of what it will take to complete the task and can envision the steps to completion. The desire to please superiors also influences people to underestimate the time to completion. Although it may not be necessary to double the time estimates, you can incorporate the most optimistic and pessimistic estimates and apply equation (1). Or, if a proven mathematical model exists, such as the KLOC estimator in equations (7) and (8), it can be used to estimate times.

- *Assign a lot of time for testing and integration.* During integration many people must work together to integrate components that may have been developed in isolation. Problems with a single component can bring the integration to a halt. Delays may be compounded by necessary re-design to correct the problem.

- *Factor in lead times for part ordering.* Even with the Internet and overnight delivery, you may find that needed parts and equipment are out-of-stock. Lead times for seemingly commonplace items can sometimes be quite lengthy.

- *Assign a project manager(s).* Consider assigning one individual who has primary responsibility for organizing and monitoring the plan. Again, the project manager must also be responsible for some of the deliverables for completion of the project.

- *Do not assign all team members to all tasks.* Experience shows that when this is the case nobody is responsible for anything and the work doesn't get done. There needs to be individual accountability for all team members. However, it may be a good idea to have more than one person responsible for activities for backup support as shown in the WBS in Table 10.1.

- *Track the progress versus the plan.* There is a tendency to create the plan and then ignore it. The plan is only valuable if it is monitored and progress is tracked.

- *Don't become a slave to the plan.* Circumstances usually dictate change. Be prepared to shift resources as needed. Monitor the plan to see if there are changes to the critical path or if a new critical path emerges.

- *Experience counts.* Get started now in developing this experience by creating a plan for your project.

10.7 Project Application: The Project Plan

A project plan should contain the following:

- *Work Breakdown Structure.* Identify the activities, deliverables, responsibilities, duration, resources, and dependencies as demonstrated in Table 10.1. Be sure to provide sufficient detail in the structure and identify clear deliverables.

- *Gantt Chart and/or Network Diagram.* Provide a graphical representation of the project plan. Network diagrams have the advantage of showing the dependencies, while Gantt charts show the timeframe. Modern software tools allow both to be integrated into the same graph and are a good compromise as demonstrated in Figure 10.3. The critical path should be identified and the float for non-critical path activities understood.

- *Costs.* Develop a tabulated list of costs and for the equipment, materials, and labor necessary to carry out the project. It may not be necessary to develop labor costs for a capstone project, yet it is good practice to estimate person-hours and compare the estimate to the actual at the end of the project.

10.8 Summary and Further Reading

Three main objectives of project management are to complete projects that are on-time, within budget, and meet the needs of the user. This chapter addressed the time and budget aspects. The key element in developing a project plan is the WBS, which is a hierarchical identification of the activities needed to complete the project. Both network diagrams and Gantt charts can be created from the WBS. A network diagram is a graphical representation of activities and their dependencies that provides for quantitative analysis of the project plan. This analysis includes computation of the critical path and float times. The Gantt chart is related to the network diagram, but provides a time scale representation of the activities. In terms of the budget issues, a simple profit and loss model was presented with a break-even analysis. Model-based techniques can be used for estimating labor costs, where the models are built from the analysis of similar projects. The models can be linear or nonlinear with single inputs. More complex models can be developed with multiple inputs in order to arrive at a more precise estimate.

Project management is a well developed field and there are many good textbooks and online resources available for delving deeper into the subject. Project Management by Gray and Larson [Gra02] is a comprehensive text on the subject that includes risk management, resource scheduling, leadership, and performance measurement. Planning, Performing, and Controlling Projects by Angus et al. [Ang00] is written for an engineering and scientific audience, and integrates phases of the design process. MindTools (www.mindtools.com) is an online resource that addresses many career skills including project management.

10.9 Problems

10.1 In your own words, describe what is meant by the work breakdown structure.

10.2 Consider the set of activities, duration (in days), and predecessors for a project given below.

Activity	A	B	C	D	E	F	G	H	I
Duration	3	9	6	6	6	3	2	6	7
Predecessors	-	-	-	A,B	D,B	C	F,E	G	F

a) Develop a network diagram representation for the project.

b) Determine the critical path.

c) Determine the float time for all activities that are not on the critical path.

10.3 Consider the set of activities, duration (in days), and predecessors for a project given below.

Activity	A	B	C	D	E	F	G	H	I	J	K
Duration	9	12	3	4	5	9	8	3	6	9	1
Predecessors	-	A	A	B,C	C	B	D	F,D	G	H,I	E

a) Develop a network diagram representation for the project.

b) Determine the critical path.

c) Determine the float time for all activities that are not on the critical path.

10.4 Explain why a network diagram cannot contain cycles. A cycle is a sequence of activities where you can travel back to an activity already visited.

10.5 Describe the advantages and disadvantages of the network diagram and Gantt chart representations for a project.

10.6 Assume that the following data has been determined for the development and sale of a new digital thermometer for home use: development cost = $250,000, production investment = $500,000, annual production volume = 20,000 units per year, and the sales lifetime is 7 years. Assuming a variable production cost of $5 per unit, determine: (a) the sales price necessary to break even within 2 years, and (b) the profit expected over the estimated sales lifetime.

10.7 Describe the difference between the cost estimation models in equations (7) and (8) and the COCOMO cost estimation model.

10.8 Consider a software development project that has a team of 50 software development engineers. The team has proposed a design and estimates that it will require

500,000 lines of code to complete the project. The average cost to the company for an engineer is $90,000 per year, including salary, benefits, and overhead. Estimate (a) the time required to complete the project, and (b) the labor costs.

10.9   Consider a software development project where the team has proposed a design and estimates that it will require 200,000 lines of code to complete. The average cost to the company for an engineer is $110,000 per year, including salary, benefits, and overhead. Estimate (a) the number of engineers needed to complete the project within 18 months, and (b) the labor costs.

10.10 **Project Application.** Develop a project plan for your project. A format and guideline for developing the plan is contained in Section 10.7.

Chapter 11 Ethical and Legal Issues

A man without ethics is a wild beast loosed upon this world.—Albert Camus

In your professional career you are going to face ethical dilemmas, and how you respond to them will affect your reputation, your employability, and the welfare of others. Systems are developed for use by other people, and the decisions that go into their design impacts the health and safety of those people. There are plenty of high profile examples, such as the two space shuttle disasters, that make it clear that the decisions of engineers and scientists have serious implications. Beyond issues in the technical aspects of design, there are professional ethics that govern the broad scope of interactions between people in the workplace. The aim of this chapter is to present the basics of engineering ethics and provide guidance for addressing dilemmas when they arise. It presents a basic overview of morality and ethics, professional codes that apply to the engineering profession, intellectual property and legal issues as they relate to design, how to apply ethics throughout the design process, and guidance for handling ethical dilemmas. The chapter concludes with case studies that provide an opportunity to apply ethical decision-making skills.

Learning Objectives

By the end of this chapter, the reader should:

- Understand what is meant by morals, ethics, and values.
- Be familiar with the IEEE Code of Ethics.
- Understand what a patent is, the criteria for filing for one, and the elements that constitute it.
- Understand the difference between patents, trade secrets, and copyrights.
- Understand the concepts of negligence and strict liability as they apply to product design.
- Understand how to incorporate ethical issues throughout the design process.
- Be able to analyze ethical case studies and suggest solutions to the dilemmas that they embody.

11.1 Ethical Theory in a Nutshell

Here is an ethical dilemma to consider as we start our discussion. Let's assume that you are conducting a job search during your senior year in college and have interviewed with several prospective employers. Company A offers a you job and you accept the position and you sign a contract agreeing upon a starting salary, position, and start date. Two weeks later, Company B also offers you a job with a higher starting salary and work that you find personally more interesting. You have a choice to make. Do you turn down the offer of Company B and stay with A? Or do you accept the offer made by B and then inform A that you are not going to work for them? This is a classic dilemma that many students will face, and rather than trying to answer it now, let's revisit it later.

There is often confusion as to what exactly is meant by the related concepts of *ethics* and *morals*. Morality is concerned with principles of right and wrong and the decisions that derive from these principles. Morals are often taught by stories that are common in all cultures, such as Aesop's fable of the boy who cried wolf. In this story, the boy was watching sheep and cried wolf when there was no wolf present. In response, all of the villagers came running to help him, only to find no wolf. When the wolf did later appear and the boy cried for help once again, nobody believed him nor came to help, and the wolf scattered the flock. The moral of this story is that it is wrong to lie, and when a person becomes known as a liar, nobody will believe him/her even if he/she is telling the truth.

The term ethics is closely related to morals and they are often used interchangeably. According to the <u>American Heritage Dictionary</u>, ethics is defined as:

> *1. Branch of philosophy that deals with the general nature of good and bad and the specific moral obligations of and choices to be made by the individual in her/his relationship to others. 2. Rules or standards governing conduct, especially those of a profession.*

Ethics is the philosophy or study of moral obligations and the choices to be made by individuals. It is important to understand that if there is no decision to be made, there is no ethical dilemma. The choices to be made are based upon a belief as to what is good or bad, and the decisions must impact relationships to others.

Morals derive from *principles,* which are fundamental laws or rules that govern behavior. An example of a principle is the Golden Rule, which states that people should treat others as they themselves would like to be treated. The Golden Rule is a universal principle and embodied in one form or another in all major religions and belief systems of the world. Another example of a principle is the belief that people should be honest and trustworthy in their dealings with others. Value is another term that is often heard in ethical discussions. A *value* is something that a person or group believes to be valuable or worthwhile. This could be relatively innocuous such as valuing baseball as a sport, or something more significant such as valuing hard work. A group of thieves may believe that it is valuable to steal from other people, but not from their own group. As these examples show, shared values may be good or bad. The final example clearly violates the principles of the Golden Rule and is considered bad by most people.

Rule-based ethics are based upon a set of rules that can be applied to make decisions. In the strictest form they are considered to be absolute in terms of governing behavior—either you follow the rule (good) or break the rule (bad). This type of an ethical system is based upon the principles of *universality* and *transitivity*. Universality means that the governing rules are such that they can be accepted by everyone, while transitivity means that you would accept others applying the same decision to you. A problem with rule-based ethics is defining a universal set of rules that everyone can agree upon. There may be a few rules or principles that can be agreed upon by all, but going beyond that is difficult.

Conditional rule-based ethics means that there are certain conditions under which an individual can break a rule. This is generally because it is believed that the moral good of the situation outweighs the rule. For example, if you have a seriously injured person in your car and are transporting them to the hospital, is it acceptable to exceed the speed limit? In this case, it may be deemed that the moral good of getting the person prompt medical attention outweighs the obligation to obey the speed limit. It is generally believed that killing others is wrong, except in the case of war. Is it acceptable to cheat on an exam because you stayed up all night to help your sick roommate? Or is it acceptable to cheat because you simply did not have enough time to study because of your other obligations? In each of these cases a moral choice has to be weighed.

In *utilitarian ethics*, decisions are made based upon the decision that brings about the highest good for all, relative to all other decisions. This sounds appealing, but has drawbacks in that bad choices may need to be made for certain parties in order to achieve this overall good. It becomes very difficult to determine exactly what the highest good is. *Situational ethics* are where decisions are made based on whether they produce the highest good for the person. In this case, decisions are made based upon the impact to the individual and situation at hand. This is generally considered a poor ethical decision-making approach.

Let's conclude this section with a new scenario. Let's assume that you are conducting a job search and Company A offers you a job, you agree upon terms of the offer, accept it, and sign a contract. You then inform the other companies that interviewed you that you are no longer available. One month later Company A calls you and notifies you that they are rescinding their offer. Although they can no longer offer you the job, they will provide you with $1,000 in compensation for rescinding the offer. You later find out that Company A just offered the exact same job to another student in your class who has a higher grade point average than yours and more experience with the technical products they design. Was the decision of the company ethical? Returning to the original case, is it ethical for the student to rescind their acceptance of the original job offer and then accept the other?

11.2 The IEEE Code of Ethics

Most engineering disciplines have associated with them a professional society or organization. In electrical and computer engineering two important ones are the Institute of Electrical and Electronics Engineers (IEEE) and the Association for Computing Machinery (ACM). The objective of professional societies is to promote and support their respective fields. They offer a variety of services and benefits to their members, such as access to technical information, networking opportunities, and financial services. They also define accepted practices of their members that are embodied in an ethical code. In fact, professional societies were originally created to provide guidance for ethical practices in their field and to ensure the safety of the public. The IEEE Code of Ethics, shown in Table 11.1, applies broadly to the electrical, computing, and software fields.

Table 11.1 The IEEE Code of Ethics.

We, the members of the IEEE, in recognition of the importance of our technologies in affecting the quality of life throughout the world, and in accepting a personal obligation to our profession, its members and the communities we serve, do hereby commit ourselves to the highest ethical and professional conduct and agree:

1. to accept responsibility in making engineering decisions consistent with the safety, health and welfare of the public, and to disclose promptly factors that might endanger the public or the environment;

2. to avoid real or perceived conflicts of interest whenever possible, and to disclose them to affected parties when they do exist;

3. to be honest and realistic in stating claims or estimates based on available data;

4. to reject bribery in all its forms;

5. to improve the understanding of technology, its appropriate application, and potential consequences;

6. to maintain and improve our technical competence and to undertake technological tasks for others only if qualified by training or experience, or after full disclosure of pertinent limitations;

7. to seek, accept, and offer honest criticism of technical work, to acknowledge and correct errors, and to credit properly the contributions of others;

8. to treat fairly all persons regardless of such factors as race, religion, gender, disability, age, or national origin;

9. to avoid injuring others, their property, reputation, or employment by false or malicious action;

10. to assist colleagues and co-workers in their professional development and to support them in following this code of ethics.

Approved by the IEEE Board of Directors, August 1990

Some of the values embodied in the code are that of treating others with fairness and respect (8–10), honesty and trustworthiness (2–4), and professional competence (6–7). The importance of the safety of the public is clearly identified in the first and ninth items of the code. The code provides a common basis for analyzing ethical cases studies and is applied for these purposes in subsequent sections.

11.3 Intellectual Property and Legal Issues

It is important to understand some legal issues, particularly as they impact product design and development. The first considered is intellectual property, which seeks to answer the question of who owns an idea or invention. The second is that of legal liability, which comes into play if somebody is injured or harmed by a product or system.

11.3.1 Intellectual Property

The intent of designing a new product is usually to sell it for a profit, which leads to the issue of ownership of both the intellectual property and the profits. The tools for protecting intellectual property are patents, trade secrets, and copyrights. Be aware that when you enter the workforce as an engineer, your employer may ask you to sign an agreement assigning their company rights to all of the intellectual property that you create while in their employ. Such contracts are common and enforceable.

The most well-known way of protecting a design or invention is with a *patent*. If a patent is held for a technology, others cannot use it without permission of the owner. The owner can deny others the right to use it, or grant the right to use it in exchange for monetary compensation. The two types of patents are utility and design patents. The United States Patent and Trademark Office (USPTO) defines the difference between the two as follows:

> *A utility patent may be granted to anyone who invents or discovers any new and useful process, machine, article of manufacture, compositions of matter, or any new useful improvement thereof. A design patent may be granted to anyone who invents a new, original, and ornamental design for an article of manufacture.*

The one that is of most interest for this discussion is the utility patent, as the design patent focuses on aesthetic design issues.

In order to be granted a utility patent, the invention must meet three conditions: it must be novel, non-obvious, and useful. Novelty means that the idea must be new and that nothing like it already exists. It cannot be an idea that is already patented or has been published for more than a year. The non-obvious condition means that another person would not be expected to develop the same idea based upon existing technology. As an example, for personal computers, it would not be possible to patent the idea of increasing data storage by adding multiple disk drives, since this is an obvious extension of existing technology. Useful means that the device must perform a useful function and be able to be reduced to practice. You could not patent the concept for a transporter of the type seen on science fiction shows, such as *Star Trek*, where people are beamed instantaneously through space from one place to another—unless of course you can reduce it to practice.

To a file a patent, extensive research needs to be done to be sure that the idea is novel. A good place to start is the US Patent Trademark Office (USPTO) website (www.uspto.gov) and its searchable database of patents back to 1790. The database allows full text search of patents from 1976 forward and patent number search back to 1790. It provides full images of all patents in the database. The website also has information on how to apply for a patent and describes the differences between patents, trademarks, and copyrights. The elements that are contained in a patent are:

- *A citation of prior art.* This is similar patents and publicly reported technology.

- *A description of the invention.* This describes how it operates and how it would be reduced to practice.

- *Claims.* They are the legal description of the invention and its unique aspects.

The claims are used to determine if another party is infringing upon a patent, and thus they must be carefully thought out and properly worded. If the claims are too broad or too specific then they may not provide much protection. There are fees for filing the patent and for periodic maintenance that can cost between $5,000 and $10,000 over the life of the patent. Hiring a good patent attorney is invaluable in the application process and will cost money—it is not unusual to spend $20,000 for lawyer and application fees. When selecting a patent attorney, make sure that he/she has a good track record and reputation.

An example patent for a hardware design of a fast floating point overflow and sign detection unit (US Patent #6,779,013) is shown in Figure 11.1. The patent is owned by Intel Corporation and the inventor was one of their employees. The first page identifies prior art that relates to this patent. This includes previous patents related to this technology, which in this case goes back to 1991. It is not unusual to cite prior patents that go back 50 or 100 years. In addition to patents, prior art in publicly available literature is identified. The first page also identifies the number of claims and supporting figures, and shows the high-level design architecture. A description of the system starts on the second page as shown in Figure 11.1 (b). The complete description is fairly lengthy and only the first page is shown. It is similar to a technical report in that it provides background on the technology and describes the invention and its operation in detail. The final page of the patent (Figure 11.1 (c)) identifies the legal claims that make the invention unique.

(12) **United States Patent**
Pangal

(10) **Patent No.:**   **US 6,779,013 B2**
(45) **Date of Patent:**   Aug. 17, 2004

(54) **FLOATING POINT OVERFLOW AND SIGN DETECTION**

(75) Inventor: **Amaresh Pangal**, Hillsboro, OR (US)

(73) Assignee: **Intel Corporation**, Santa Clara, CA (US)

(*) Notice: Subject to any disclaimer, the term of this patent is extended or adjusted under 35 U.S.C. 154(b) by 368 days.

(21) Appl. No.: **09/873,744**

(22) Filed: **Jun. 4, 2001**

(65) **Prior Publication Data**

US 2002/0194239 A1 Dec. 19, 2002

(51) Int. Cl.[7] **G06F 7/44**
(52) U.S. Cl. **708/503**
(58) Field of Search 708/503, 501

(56) **References Cited**

U.S. PATENT DOCUMENTS

5,764,089 A	6/1998	Partovi et al.	327/200
5,898,330 A	4/1999	Klass	327/210
5,900,759 A	5/1999	Tam	327/201
5,993,051 A *	11/1999	Jiang et al.	708/501
6,205,462 B1 *	3/2001	Wyland et al.	708/503
6,360,189 B1 *	3/2002	Hinds et al.	708/501
6,480,872 B1 *	11/2002	Choquette	708/501

OTHER PUBLICATIONS

Beaumont–Smith, A., et al., "Reduced Latency IEEE Floating–Point Standard Adder Architectures", *Proceedings of the 14th IEEE Symposium on Computer Arithmetic*, 8 pgs., (1998).
Even, G., et al., "On the Design of IEEE Compliant Floating Point Units", *IEEE Transactions on Computers*, vol. 49, 398–413, (May 2000).
Goto, G., et al., "A 54×54–b Regularly Structured Tree Multiplier", *IEEE Journal of Solid–State Circuits*, vol. 27, 1229–1236, (Sep. 1992).

Ide, N., et al., "2.44–GFLOPS 300–MHz Floating–Point Vector–Processing Unit for High–Performance 3–D Graphics Computing", *IEEE Journal of Solid–State Circuits*, vol. 35, 1025–1033, (Jul. 2000).
Klass, F., "Semi–Dynamic and Dynamic Flip–Flops with Embedded Logic", *Proceedings of the Symposium on VLSI Circuits, Digest of Technical Papers*, Honolulu, HI, IEEE Circuits Soc. Japan Soc. Appl. Phys. Inst. Electron., Inf. & Commun. Eng. Japan, pp. 108–109, (1998).
Lee, K.T., et al., "1 GHz Leading Zero Anticipator Using Independent Sign–Bit Determination Logic", *2000 Symposium on VLSI Circuits Digest of Technical Papers*, 194–195, (2000).
Partovi, H., et al., "Flow–Through Latch and Edge–Triggered Flip–Flop Hybrid Elements", *Proceedings of the IEEE International Solid–State Circuits Conference, Digest of Technical Papers and Slide Supplement*, NexGen Inc., Milpitas, CA, 40 pgs., (1996).

(List continued on next page.)

Primary Examiner—Tan V. Mai
(74) *Attorney, Agent, or Firm*—Schwegman, Lundberg, Wosessner & Kluth, P.A.

(57) **ABSTRACT**

A multiply-accumulate circuit includes a compressor tree to generate a product with a binary exponent and a mantissa in carry-save format. The product is converted into a number having a three bit exponent and a fifty-seven bit mantissa in carry-save format for accumulation. An adder circuit accumulates the converted products in carry-save format. Because the products being summed are in carry-save format, post-normalization is avoided within the adder feedback loop. The adder operates on floating point number representations having exponents with a least significant bit weight of thirty-two, and exponent comparisons within the adder exponent path are limited in size. Variable shifters are avoided in the adder mantissa path. A single mantissa shift of thirty-two bits is provided by a conditional shifter.

20 Claims, 10 Drawing Sheets

Figure 11.1 (a) Example of a US patent for a floating point overflow and sign detection unit.

US 6,779,013 B2

| 1 | 2 |

FLOATING POINT OVERFLOW AND SIGN DETECTION

FIELD

The present invention relates generally to floating point operations, and more specifically to floating point multiply accumulators.

BACKGROUND

Fast floating point mathematical operations have become an important feature in modern electronics. Floating point units are useful in applications such as three-dimensional graphics computations and digital signal processing (DSP). Examples of three-dimensional graphics computation include geometry transformations and perspective transformations. These transformations are performed when the motion of objects is determined by calculating physical equations in response to interactive events instead of replaying prerecorded data.

Many DSP operations, such as finite impulse response (FIR) filters, compute $\Sigma(a_i b_i)$, where i=0 to n−1, and a_i and b_i are both single precision floating point numbers. This type of computation typically employs floating point multiply accumulate (FMAC) units which perform many multiplication operations and add the resulting products to give the final result. In these types of applications, fast FMAC units typically execute multiplies and additions in parallel without pipeline bubbles. One example FMAC unit is described in: Nobuhiro et al., "2.44-GFLOPS 300-MHz Floating-Point Vector Processing Unit for High-Performance 3-D Graphics Computing," IEEE Journal of Solid State Circuits, Vol. 35, No. 7, July 2000.

The Institute of Electrical and Electronic Engineers (IEEE) has published an industry standard for floating point operations in the ANSI/IEEE Std 754–1985, *IEEE Standard for Binary Floating-Point Arithmetic*, IEEE, New York, 1985, hereinafter referred to as the "IEEE standard." A typical implementation for a floating point FMAC compliant with the IEEE standard is shown in FIG. 1. FMAC **100** implements a single precision floating point multiply and accumulate instruction "D=(A×B)+C," as an indivisible operation. As can be seen from FIG. 1, fast floating point multipliers and fast floating point adders are both important ingredients to make a fast FMAC.

Multiplicands A and B are received by multiplier **110**, and the product is normalized in post-normalization block **120**. Multiplicands A and B are typically in an IEEE standard floating point format, and post-normalization block **120** typically operates on (normalizes) the output of multiplier **110** to make the product conform to the same format. For example, when multiplicands A and B are IEEE standard single precision floating point numbers, post-normalization block **120** operates on the output from multiplier **110** so that adder **130** receives the product as an IEEE standard single precision floating point number.

Adder **130** adds the normalized product from post-normalization block **120** with the output from multiplexer **140**. Multiplexer **140** can choose between the number C and the previous sum on node **152**. When the previous sum is used, FMAC **100** is performing a multiply-accumulate function. The output of adder **130** is normalized in post-normalization block **150** so that the sum on node **152** is in the standard format discussed above.

Adder **130** and post-normalization block **150** can be "non-pipelined," which means that an accumulation can be performed in a single clock cycle. When non-pipelined, adder **130** and post-normalization block typically include sufficient logic to limit the frequency at which FMAC **100** can operate, in part because floating point adders typically include circuits for alignment, mantissa addition, rounding, and other complex operations. To increase the frequency of operation, adder **130** and post-normalization block **150** can be "pipelined," which means registers can be included in the data path to store intermediate results. One disadvantage of pipelining is the introduction of pipeline stalls or bubbles, which decrease the effective data rate through FMAC **100**.

For the reasons stated above, and for other reasons stated below which will become apparent to those skilled in the art upon reading and understanding the present specification, there is a need in the art for fast floating point multiply and accumulate circuits.

BRIEF DESCRIPTION OF THE DRAWINGS

FIG. 1 shows a prior art floating point multiply-accumulate circuit;

FIG. 2 shows an integrated circuit with a floating point multiply-accumulate circuit;

FIG. 3 shows the exponent and mantissa paths of a floating point multiply-accumulate circuit;

FIG. 4 shows a mantissa multiplier circuit;

FIG. 5 shows a floating point conversion unit;

FIG. 6 shows a carry-save negation circuit;

FIG. 7 shows a base 32 floating point number representation;

FIG. 8 shows an exponent path of a floating point adder;

FIG. 9 shows a mantissa path of a floating point adder;

FIG. 10 shows an overflow detection circuit;

FIG. 11 shows a post-normalization circuit; and

FIG. 12 shows a sign detection circuit.

DESCRIPTION OF EMBODIMENTS

In the following detailed description of the embodiments, reference is made to the accompanying drawings which show, by way of illustration, specific embodiments in which the invention may be practiced. In the drawings, like numerals describe substantially similar components throughout the several views. These embodiments are described in sufficient detail to enable those skilled in the art to practice the invention. Other embodiments may be utilized and structural, logical, and electrical changes may be made without departing from the scope of the present invention. Moreover, it is to be understood that the various embodiments of the invention, although different, are not necessarily mutually exclusive. For example, a particular feature, structure, or characteristic described in one embodiment may be included within other embodiments. The following detailed description is, therefore, not to be taken in a limiting sense, and the scope of the present invention is defined only by the appended claims, along with the full scope of equivalents to which such claims are entitled.

Floating Point Multiply Accumulator

FIG. 2 shows an integrated circuit with a floating point multiply-accumulate circuit. Integrated circuit **200** includes floating point multiplier **210**, floating point conversion unit **220**, floating point adder **230**, and post-normalization circuit **250**. Each of the elements shown in FIG. 2 is explained in further detail with reference to figures that follow. In this section, a brief overview of the FIG. 2 elements and their

Figure 11.1 (b) First page of the patent description.

US 6,779,013 B2

| 11 | | 12 |

-continued

S1	C1	MC	Sign
1	0	0	0
1	0	1	1
1	1	X	-

Magnitude comparator **325** operates in parallel with adder mantissa path **324**, so MC is available for sign detection circuit **1104** at substantially the same time as Msum. In this manner, the operation of sign detection circuit **1104** does not appreciably increase the delay within the feedback loop.

CONCLUSION

The method and apparatus of the present invention provide a fast multiply-accumulate operation that can be made compliant with any floating point format. Furthermore, the method and apparatus of the present invention can provide precision comparable to the precision available using prior art double precision arithmetic units, in part because the mantissa fields are expanded. In some embodiments, IEEE standard single precision operands are multiplied and the products are summed. The multiplier includes a compressor tree to generate a product with a binary exponent and a mantissa in carry-save format. The product is converted into a number having a three bit exponent and a fifty-seven bit mantissa in carry-save format for accumulation. An adder circuit accumulates the converted products in carry-save format. Because the products being summed are in carry-save format, post-normalization is avoided within the adder feedback loop. In addition, because the adder operates on floating point number representations having exponents with a least significant bit weight of thirty-two, exponent comparisons within the adder exponent path are fast, and variable shifters can be avoided in the adder mantissa path. When the adder is not pipelined, a fast single cycle accumulation is realized with the method and apparatus of the present invention.

It is to be understood that the above description is intended to be illustrative, and not restrictive. Many other embodiments will be apparent to those of skill in the art upon reading and understanding the above description. The scope of the invention should, therefore, be determined with reference to the appended claims, along with the full scope of equivalents to which such claims are entitled.

What is claimed is:

1. A floating point multiply-accumulate circuit comprising:

an exponent path including:

an exponent summer to sum two input exponents having a first weight to produce a product exponent;

an exponent conversion unit coupled to the output of the exponent summer, to convert the product exponent to a second weight; and

an exponent accumulation stage to accumulate the converted product exponent and to choose a larger exponent from the converted product exponent and an accumulated exponent; and

a mantissa path including:

a mantissa multiplier to multiply two input mantissas and produce a product mantissa in carry-save format that includes a sum field and a carry field;

a mantissa shifter to shift the sum field and the carry field responsive to the exponent conversion unit in the exponent path; and

a mantissa accumulator to accumulate shifted product mantissas in carry-save format, the mantissa accumulator including an overflow detection circuit responsive to two most significant bits of a sum field output from the mantissa accumulator.

2. The floating point multiply-accumulate circuit of claim 1 wherein the overflow detection circuit comprises an exclusive-or gate.

3. The floating point multiply-accumulate circuit of claim 1 wherein the exponent conversion unit is configured to zero the least significant five bits of the product exponent.

4. The floating point multiply-accumulate circuit of claim 1 wherein the mantissa shifter is configured to shift the sum field and carry field of the product mantissa by a number of bit positions equal to a value of the least significant five bits of the product exponent.

5. The floating point multiply-accumulate circuit of claim 1, wherein the mantissa accumulator comprises four-to-two compressors.

6. The floating point multiply-accumulate circuit of claim 1 further comprising a post-normalization stage to produce a normalized floating point resultant.

7. The floating point multiply-accumulate circuit of claim 6 wherein the post-normalization stage includes a sign detection circuit.

8. The floating point multiply-accumulate circuit of claim 7 further comprising a magnitude comparator in parallel with the mantissa accumulator, wherein the sign detection circuit is responsive to the magnitude comparator.

9. The floating point multiply-accumulate circuit of claim 6 wherein the post-normalization stage is configured to be turned off until accumulation is complete.

10. The floating point multiply-accumulate circuit of claim 1 wherein the exponent conversion unit is configured to convert the product exponent to have a least significant bit weight equal to thirty-two.

11. A method performed within a programmable digital computer, comprising:

multiplying two floating point mantissas and summing two floating point exponents to form a product;

converting the product to have a different least significant bit weight exponent field;

accumulating the converted product in carry-save format;

detecting overflow as a function of two most significant bits of a sum field of an accumulated product; and

post-normalizing the accumulated product.

12. The method of claim 11 wherein accumulating the product comprises adding a first plurality of products with a last product, the method further comprising turning off post-normalization until the last product is accumulated.

13. The method of claim 11 wherein converting comprises:

shifting a mantissa of the product by an amount equal to the value of the least significant five bits of the exponent of the product; and

zeroing the least significant five bits of an exponent of the product.

14. The method of claim 11 wherein accumulating comprises:

comparing an exponent of a first converted product to an exponent of a second converted product;

conditionally shifting right by a fixed amount the mantissa of the converted product having a smaller exponent;

selecting the larger exponent as a resultant exponent; and

producing a resultant mantissa from a mantissa of the first converted product and a mantissa of the second converted product.

Figure 11.1 (c) Conclusion of the patent and its claims.

In the United States, patents are granted based upon the concept of first to conceive the invention, not first to file the patent application. If two parties are applying for a patent at the same time, then the one who demonstrates that they were the first to conceive the idea and reduce it to practice receives the patent. Good records must be maintained to prove this. It is done by recording inventions in a bound design notebook with numbered pages that cannot be removed. All entries should be clearly described, understandable by other engineers, and be signed and dated. For the best protection, entries should be signed and witnessed by at least one other person and the notebook should be occasionally notarized. Pages should not be removed from the notebook and entries should be made in non-erasable pen. Mistakes and blank spaces should be crossed out, signed, and dated. Figures should be drawn directly in the notebook, while computer generated figures are pasted in so that they are not removable. It is good practice to maintain a design notebook even if you are not planning to patent your ideas.

Once a patent is granted, it is good for 20 years—after it expires, the invention is fair game for anybody to use. Let's assume that you are the owner of a patent and you find that somebody is infringing upon it. Is the government going to come in and start legal action against the infringer? No—a patent only gives the owner a right to sue others if they infringe upon it. The government does not actively seek out offenders and protect your intellectual property. The owner of the patent must protect it.

There are drawbacks to patents. The owner must be vigilant in defending a patent and may have to initiate legal action to do so. Once an idea is patented, it is made public for all to see on the government's website. This may not be a good idea depending upon the technology and competitive situation. An alternative that many companies employ to protect intellectual property is to hold them as *trade secrets*. Obviously, the idea must be kept secret so that others cannot find out about it. This is often done by restricting the number of people who have access to the idea and by having those who do know about it sign a *non-disclosure agreement*. It is common practice for companies to ask employees and visitors to their facilities to sign a non-disclosure agreement that prevents the signer from disseminating information about their products, services, and trade secrets. If one breaks a non-disclosure agreement, they can be held legally liable. Trade secrets pose another type of risk to a company, since once the secret is revealed, it is fair game for competitors. One way to determine a competitor's trade secrets is through the use of *reverse-engineering*, where a device or process is taken apart to understand how it works. Reverse-engineering is legal if the information is obtained through legal means, but one must be careful with the information that is obtained and not copy it unless legally allowed to do so. The Digital Millennium Copyright Act of 2000 prohibits breaking technological protections, such as encryption, to learn about a competitor's product.

Copyrights protect published works such as books, articles, music, and software. A copyright means that others cannot distribute copyrighted material without permission of the owner. It is easy to obtain a copyright—all that a person has to do to copyright material is indicate the word "copyright" on the work, along with the year of publication, and identify the name of the copyright holder. Copyrights can be officially registered through the US Copyright Office, and it is a good idea to do so for involved works such as writing a book or publishing music. Registering provides a stronger legal basis for pursuing claims. Copyrights are good for the lifetime of the holder plus 50 years, while for a company they are good for 75 years.

11.3.2 Liability and Negligence

Civil lawsuits, which are those that are relevant for our discussion, are those in which one party sues another. They involve something that is known as a *tort,* which serves as grounds for the lawsuit. A tort is a wrongful act, though not necessary illegal, for which a civil lawsuit can be brought, including product liability. A company or person can be sued for damages caused by a product design and be held *liable* for them, meaning required to pay monetary damages.

The two standards for determining legal liability in tort law are that of negligence and strict liability. In the case of *negligence,* it must be shown that the manufacturer did not follow reasonable standards and rules that apply to the situation and also committed a wrongful act. Exactly what constitutes reasonable has to be determined for the particular case. For a product design, a manufacturer can be held legally liable for negligence if the plaintiff demonstrates that the following four conditions hold true:

1. The manufacturer had a duty to follow reasonable standards and rules.
2. There was a breach of duty (i.e., failed to include safety devices).
3. The plaintiff was harmed.
4. The breach caused the harm.

Depending on the severity of the danger, there are different levels of negligence: simple, gross, and criminal. Negligence claims can be brought for design flaws, manufacturing defects, and for failing to warn the user of safety hazards.

An even less stringent standard than negligence is known as *strict liability,* in which the following four conditions must hold true:

1. The product was dangerous and/or defective.
2. The defect existed when it left the manufacturer's control.
3. The defect caused harm.
4. The harm is assignable to the defect.

Strict liability focuses only on the product itself—if the product contains a defect that caused harm, the manufacturer is liable. This is regardless of whether there was negligence, if safety devices were incorporated, or if the user was warned of potential dangers. If the product had a defect or was dangerous when it left the hands of the manufacturer, the manufacturer may be liable.

11.4 Handling Ethical Dilemmas

You are going to encounter ethical dilemmas in your career, both technical and based upon interpersonal relations. Ethical dilemmas are not always obvious and can be quite subtle. A supervisor probably won't say *"Could you falsify this data for me so we can ship the product?"* The pressure will likely be much more subtle and over the course of multiple conversations, unlike Dilbert's dilemma in Figure 11.2. Consider the following sequence of statements: *"We have invested a lot of time and money in the design."; "We really need this system to work."; "The company's future depends upon this."; "Is there any way that we can make adjustments to make it pass the certification."; "Is it close enough that we could certify it? It really meets the needs and the standard has a margin of error built in to it."*

DILBERT® by Scott Adams

Figure 11.2 Dilbert's ethical dilemma. (Dilbert © United Feature Syndicate. Reprinted by permission.)

A framework for considering decisions is shown in the 2x2 matrix shown in Figure 11.3 [Die00]. The idea is that decisions have both ethical and legal dimensions to them. The legality may be in terms of internal company policies or local, state, and federal laws. Quadrant I decisions are clearly to be avoided as they are not legal or ethical. Quadrant II decisions present an interesting dilemma since they are legal, but yet are not ethical. Making such a decision may not have punitive ramifications, but will have a negative impact on your professional reputation. Quadrant III decisions certainly feel right and are tempting, since they are moral but not legal. It is in Quadrants II and III where most moral dilemmas take place. Taken together, II and III represent opportunities for reform, to change the system positively so that ethical choices are legal and unethical choices are illegal. Left uncorrected, both lead to cynicism with the system, whether it is company policy or the legal system. Correcting them may take longer to address than the immediate dilemma, but have the potential for high payoff. Clearly, quadrant IV decisions are best and the goal in the decision process.

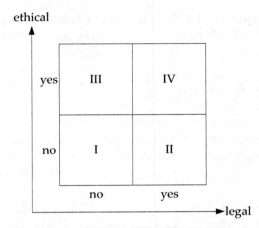

Figure 11.3 A 2x2 matrix for ethical decision making [Die00].

One way to evaluate a decision is through what is known as the *newspaper test*. The idea is to ask whether you would be comfortable if the decision were published in a newspaper for all to see. Advice from trusted friends or colleagues can be valuable, but if they are impacted by the decision, they may not be a good source. Many companies have an ethics office that provides guidance to which employees can turn. The IEEE and Online Ethics Center for Engineering and Science co-sponsor an ethics help-line at www.OnlineEthics.org, where ethical questions can be submitted.

There may be situations in which you are put in an ethical dilemma by your employer that cannot be resolved internally. As a last resort, you may go outside of the company to the press or a government agency to report the problem. This is known as being a *whistleblower*. An example of this is the space shuttle Challenger accident, in which the explosion was caused by O-ring failures in the booster rockets made by Morton-Thiokol. The engineers and scientists recommended against the launch, but their superiors at NASA and Morton-Thiokol went ahead over their objections. The engineers later "blew the whistle" by going public with this information. Although it was too late to avert the disaster, it was valuable in revealing what had happened and preventing future mishaps. The following criteria should be satisfied when considering becoming a whistleblower [Deg81]:

- The harm to the public must be considerable or serious.

- Concerns must have been made to your superiors (up to the CEO) without satisfactory resolution.

- You have documented evidence that would convince an impartial observer that your company is wrong. You should have clear technical information to support your claims.

- Release of the information outside of the company will prevent the harm.

Whistleblowing does have risks, particularly loss of job, but the risks may be well worth it. The Sarbanes-Oxley Act of 2002 improved federal protection of whistleblowers in the wake of corporate scandals in the 1990s.

11.5 Case Study Analysis

A good way to develop decision-making skills is by examining case studies. We now go through the steps of analyzing a case study, employing the IEEE Code of Ethics as a guide. In order to do so, we apply a paradigm that is a modification of the one used by Lockheed Martin Corporation in their employee ethics training programs [Loc97]. The modified paradigm is as follows:

1. *Gather information.* What things are known about the situation, but also very importantly, what isn't known, and what assumptions are being made?

2. *Identify the stakeholders.* Who will be affected by the decision? They may include you, your company, your supervisors, your professional colleagues, your profession, the public, and users of your products.

3. *Consider what ethical values are relevant to this situation.* Identify the elements of the IEEE Code of Ethics and legal issues that apply to the situation.

4. *Determine a course of action.* Identify different alternative decisions and actions. Select the one that you believe best meets the interests of the stakeholders and the ethical values.

Example 11.1 demonstrates the application of this paradigm to a case study for a flawed chip design. More case studies are provided in the end of chapter problems.

Example 11.1 Ethics case study for a flawed chip design. (Texas A&M Ethics Case studies, http://ethics.tamu.edu. Reprinted by permission.)

Your company has designed a chip for a new scientific calculator that features high-precision floating-point accuracy to 17 significant digits for all 250 mathematical functions provided. After one-and-a-half years in development, and after shipping over 500 beta units to key customers, the company discovers a problem with certain calculations. In addition, the company has manufactured 5000 more calculators with this chip that are ready to be shipped.

In order to expedite floating point operations (used in handling scientific notation in mathematical operations) in a computer or calculator, certain tables of values often are used to assist in the speed of execution. (For example, a calculator requiring as long as 3 minutes to perform a tangent calculation would have no market appeal.) These tables can contain up to 100 integer entries. During beta testing, you discover that several of these values were incorrectly entered before burning them into the firmware.

Further testing concludes that because of the location and use of these table errors, the only mathematical results affected will occur in the 13th to the 17th significant digits for the double precision floating point operations. Your management is applying subtle pressure to release the chips due to the time and money invested in the project so far. Identify a plan of action.

Discussion:

Step 1: Gather information

The information makes it clear that the calculator may fail if precision beyond the 13th digit is needed. What is not known is the standard precision of a calculator. An Internet search of calculator specifications was conducted, and it was found that most scientific calculators have a 10 or 12-digit display plus two digits for the power. Going to the 13th digit implies a higher than normal precision calculator. A calculator with 17 significant digits would likely be used by scientists and engineers for high precision calculations.

Step 2: Identify stakeholders

The following are possible stakeholders:

- Users of the calculator. In certain situations, the user can make an incorrect calculation.
- Public. The users of calculator may perform calculations that could impact safety of public. This could be a real possibility given the likely users.
- Company and employees. There are negative ramifications of releasing a faulty product. This could result in monetary harm to the employees of the company that you work for.

Step 3: Identify relevant ethical values

Values from the IEEE Code are identified with a discussion of why they apply:

1: Release of the faulty calculator has the potential to endanger the safety of the public.

3: Need to be honest in stating claims as to the precision of the calculator.

9: There is clear potential to injure the reputation of the company and its employees.

In terms of legal issues, the company would be opening itself to claims of negligence, particularly since the defect was identified prior to release of the product.

Step 4: Determine a course of action

Three possible courses of action are:

A: Release the product as is without notifying the customers. This is not a good choice due to the potential harm to the safety of the public. This would not pass the newspaper test. It also may be illegal if the calculator is advertised to work to 17 significant digits.

B: Use the chips in a different calculator that is only guaranteed to compute at a lower precision, if such an option exists. The company would have to be producing one, and the technology would have to be compatible.

C: Throw away the chips and take a loss on their production and correct the problem. Conduct testing again to verify that the corrected chips work properly.

Options B and C are both reasonable choices. Option B reduces economic losses, but acceptance testing should be conducted to make sure they operate properly at the lower precision. Option C is the safest solution. Option A might be seen as a viable choice, however, since it could be reasoned that handheld calculators are not often used for safety critical applications. That is a risky assumption and the potential negative effects are great. Also note that the ideal situation would have been to develop a test plan (Chapter 7) that would have caught this error before release.

11.6 Project Application: Incorporating Ethics in the Design Process

There are good design practices that can be followed to minimize the chances that a product will be unsafe. The first and most important issue is to identify safety and health factors that impact the system and address them in the design. Its importance is reflected clearly as the first item in the IEEE Code of Ethics. It is a fundamental canon in virtually all professional engineering ethical codes. Be aware that there are often tradeoffs between economic and safety constraints and one may be satisfied at the expense of the other. It is also important to follow good practices throughout the design process. Ways that ethical considerations are included throughout the design process are:

- *Conduct research so that prior art is understood.* This will minimize conflicts over intellectual property ownership.
- *Make sure that the requirements meet the needs of the stakeholders.* If they do not, the wrong system will be designed. This is the concept of validation that was addressed in Chapter 3. Be honest and realistic in making claims about what the system can do and realize that all have limitations. There is often tension between engineering and marketing functions of an organization, with engineers often believing that marketing is overstating the capabilities of the product. A properly developed Requirements Specification is an effective tool for communicating the proposed system functionality to all parties.

- *Identify and apply safety standards.* There are many codes and guidelines that address safety in design for particular technologies. They should be identified and applied as necessary. For example, it may be wise to design a consumer product to meet UL (Underwriter's Laboratory) guidelines. It may difficult to fully incorporate guidelines in a capstone design project, depending upon the complexity of the design and the standards. However, consideration should be given to identifying those that apply.

- *Keep the design space large and explore as many solutions as possible.* The hallmark of design is that there is typically no single solution to the problem, but many that may **satisfice**. Satisfice means that a solution may meet the design requirements, but may not be the optimal solution. It is important to fully understand the technical tradeoffs involved in a given design and their impact on health and safety.

- *Consider all of the possible ways that a system can fail.* This follows the IEEE Code of Ethics, which indicates that engineers should understand technology, its application, and potential consequences. This can be done formally through the use of techniques such as failure mode analysis and the examination of system reliability (Chapter 8). In safety critical systems, the use of redundant systems may be warranted to improve reliability. An excellent example of a technical design that incorporates redundancy in a safety critical application for a hot tub controller is found in the article *Designing for Reliability, Maintainability, and Safety* by George Novacek in <u>Circuit Cellar</u> magazine [Nov00, Nov01].

- *Identify ways in which the system can fail by misuse or operator error.* This requires thinking like a beginner who is completely unfamiliar with the technology. Provide manuals for operation and safety labels where appropriate.

- *Make realistic cost and project schedule estimates.* The IEEE Code of Ethics indicates that engineers should be honest and realistic in stating claims, and this applies to the project plan as well as the system cost.

- *Conduct design reviews.* The purpose of a design review is to have peers who are knowledgeable about the design and the technology conduct a review of the work you have done. This can be a humbling, but valuable, experience. It is also in keeping with the IEEE Code of Ethics, which indicates that engineers should seek, accept, and offer honest criticism of technical work and acknowledge and correct errors.

- *Verify the engineering requirements during testing.* We saw in Chapter 3 that verification is the process of showing that the system is built properly. Verification usually occurs at the end of a project at a time that there may be a great deal of pressure to complete the tests and finish the project.

11.7 Summary and Further Reading

This chapter provided a brief overview of ethics, morality, and different ethical decision making systems. The IEEE Code of Ethics was presented, which serves as a set of ethical values for the electrical and computer engineering profession and provides a basis for analyzing ethical dilemmas. Legal issues surrounding intellectual property and product liability were provided. Advice for incorporating safety and ethics in design projects was provided as well as a paradigm for analyzing case studies.

The background on ethical frameworks and its importance in engineering came from a number of articles. The article *What has Ethics to do with me?: I am an Engineer* [Gob99] provides background on rule-based ethics, universality, and transitivity. *A Piercian Approach to Professional Ethics Instruction* [Cha02b] addressed ethical decision-making frameworks. *Ethical Considerations in the Engineering Design Process* [Van01] emphasizes the need to keep the design space as large as possible. The article *Integrating Ethics and Design* [Mcl93] does a good job of describing three different levels of ethical decision making in technical, professional, and social contexts. The two engineering design textbooks by Dieter [Die00] and Hyman [Hym98] are good resources on legal and intellectual property issues. The United States Patent and Trademark Office has an excellent website (www.uspto.gov) with information on patents, applying for a patent, and a searchable patent database. The website www.HowStuffWorks.com is also a good source on ethical theory, patents, and legal issues surrounding product liability.

11.8 Problems

11.1 Describe the relationship between ethics and morals.

11.2 Describe the differences between morals and values.

11.3 Which patent is most relevant for engineering inventions, a design patent or utility patent? Why?

11.4 What are the criteria that are used in evaluating patents?

11.5 Explain the importance of claims in a patent application.

11.6 Discuss the tradeoffs involved between using patents and trade secrets to protect intellectual property.

11.7 When can reverse-engineering be used, and how can the information obtained from it be used?

11.8 What is the difference between negligence and strict liability in tort law?

11.9 For the case study presented below, apply the ethical decision making paradigm presented in Section 11.5 to analyze the situation. Present potential solutions to the scenario and provide a discussion of them.

Case Study: Disk Drive Diagnostics. (Copyright John Wallberg. Reprinted by permission.)

SCSI, an industry standard system for connecting devices (like disks) to computers, provides a vendor ID protocol by which the computer can identify the supplier (and model) of every attached disk.

Company C makes file servers consisting of a processor and disks. Disks sold by C identify C in their vendor ID. Disks from other manufacturers can be connected to C's file servers; however, the file server software performs certain maintenance functions, notably pre-failure warnings based on performance monitoring, only on C-supplied disks.

Company P decides to compete with C by supplying cheaper disks for C's file server. They quickly discover that while their disks work on C's file servers, their disks lack a pre-failure warning feature that C's disks have. Therefore, the CEO of P directs you, the engineer in charge of the disk product, to find a solution to the problem of no pre-failure warning for your disks. Using reverse engineering, you discover that by changing the vendor ID of P's disks, the C file servers will treat P disks as C disks. Your management at company P instructs you to incorporate this change into your product so that you can advertise the disks as "100% C-compatible." What would you do in this situation?

11.10 For the case study presented below, apply the ethical decision making paradigm presented in Section 11.5 to analyze the situation. Present potential courses of action and provide a discussion of them.

Case Study: Encryption Software (Texas A&M Ethics Case Studies, http://ethics.tamu.edu. Reprinted by permission.)

You are a recently hired engineer who has been recruited directly out of college. For your first assignment, your boss asked you to write a piece of software to provide security from "prying eyes" over emailed documents; these documents would be used internally by the company. This software will subsequently be distributed to different departments.

Upon completion of this software project, you saw a program on the local news about an individual in California who has made similar software available overseas. This individual is currently under prosecution in a federal court for the distribution of algorithms and information which (by law) must remain within the United States for purposes of national security.

It occurs to you that your company is a multinational corporation and that the software might have been distributed overseas. You then discover that the software has indeed been sent overseas to other offices within the corporation. You speak with your boss, informing him of the news program from the night before. He shrugs off this comment, stating that "The company is based in the United States and we are certainly no threat to national security in any way. Besides, there's no way anyone will find out about software we use internally."

You agree with your boss, and let it go. Later on however, you receive a letter from a gentleman working as a contractor for his company overseas. Through some correspondence regarding the functionality of the software and technical matters, you learn that the Middle Eastern office had been supplying his software outside the company to contractors and clients so that they could exchange secure emailed documents. What would you do in this situation?

11.11 For the case study presented below, apply the ethical decision making paradigm presented in Section 11.5 to analyze the situation. Present potential courses of action and provide a discussion of them.

Case Study: A Failure. (Texas A&M Ethics Case Studies, http://ethics.tamu.edu. Reprinted by permission.)

You work for Velky Measurement which has for years provided DGC Corporation with sophisticated electronic equipment for patient health monitoring systems. Recently, DGC returned a failed piece of measurement equipment. A meeting was held with representatives of Velky and DGC to discuss the problem. This included you and your project manager who is intimately acquainted with the returned equipment. During the course of the meeting it becomes apparent to you that the problem has to be Velky's. You suspect that the equipment failed because of an internal design problem and that it was not properly tested. However, at the conclusion of the meeting your project manager represents Velky's official position—the test equipment is functioning properly.

You keep silent during the meeting, but afterwards talk to your project manager about his diagnosis. You suggest that Velky tell DGC that the problem is due to a design fault and that Velky will replace the defective equipment. You manager replies, "*I don't think it's wise to acknowledge that it's our fault. There's no need to hang out our wash and lessen DGC's confidence in the quality of our work. A good will gesture to replace the equipment should suffice.*"

Utlimately, Velky's management replaces the equipment because DGC has been such a good customer. Although Velky replaces the equipment at its own expense, it does not disclose the real nature of the problem. What would you do in this situation?

11.12 For the case study presented below, apply the ethical decision making paradigm presented in Section 11.5 to analyze the situation. Present potential courses of action and provide a discussion of them.

Case Study: A Vacation (Texas A&M Ethics Case Studies, http://ethics.tamu.edu. Reprinted by permission.)

You work for Rancott and were looking forward to an upcoming trip for weeks. Once you were assigned to help install Rancott's equipment for Boulding Corporation, you arranged a vacation at a nearby ski resort. The installation was scheduled to be completed on the 12th and your vacation would begin on the 13th. That meant a full week of skiing with three of your old college buddies.

Unfortunately, not all of the equipment arrived on time. Eight of the ten identical units were installed by mid-morning on the 12th. Even if the remaining two units had arrived that morning, it would take another full day to install them. However, you were informed that it might take as long as two more days for the units to arrive.

"*Terrific,*" you sighed, "*there goes my vacation—and all the money I put down for the condo.*" "*No problem,*" replied Jerry, the Boulding engineer who had worked side-by-side with you as each of the first eight units was installed. He said "*I can handle this for you. We did the first eight together. It's silly for you to have to hang around and blow your vacation.*" Jerry knew why you were sent to supervise the installation of the new equipment. It had to be properly installed in order to avoid risking injuries to those who use it. Although you are aware of this, you are confident that Jerry is fully capable to supervise the installation of the remaining two units. What would you do?

Chapter 12 Oral Presentations

Nothing should be explained in a way that it cannot be understood by an intelligent 12 year old. — Albert Einstein

We can all probably remember the anticipation of our first oral presentation—the sweaty palms, butterflies in the stomach, and the pressure of trying to remember all the points to be made. Then there is the fear associated with standing up in front of peers or teachers and presenting ideas to have them openly criticized. According to a 1973 <u>London Times</u> survey, Americans are more afraid of speaking in front of groups than dying. Perhaps this is due to the fact that although people know they will die someday, the danger associated with giving a presentation is more imminent and a greater concern. Somewhere in the capstone design experience, it is likely that you will have to make an oral presentation. Examples are the project proposal, a mid-term design review, and the final presentation. The ability to communicate your ideas is important beyond your academic career, since practicing engineers are often required to make oral presentations. Further, your overall ability to communicate influences how others will accept your ideas and act upon them—those who communicate clearly are held in high regard by their peers and tend to advance more quickly. The good news is that there is help to overcome the fear of oral presentations. With practice and adherence to some basic principles, one can become a competent, if not excellent, communicator.

Learning Objectives

By the end of this chapter, the reader should:

- Understand how people evaluate oral presentations.
- Understand common elements of a technical presentation.
- Be able to assemble an effective presentation.

DILBERT® by Scott Adams

© 2003 United Feature Syndicate, Inc.

Figure 12.1 PowerPoint disability. (Dilbert © United Feature Syndicate. Reprinted by permission.)

12.1 How People Evaluate Presentations

It is informative to understand how your audience responds to and evaluates oral presentations. Listening to a presentation is strongly associated with what is referred to as right-brain activity. Right-brain activity is dominated by emotion and intuition, while left-brain activity is associated with logical thinking and reason. This is an oversimplified model of the brain, but the point is that emotion and intuition are important elements that people rely on when evaluating a presentation.

There are three elements, known as the "three V's," that constitute a presentation: the verbal, the vocal, and the visual. Verbal is what the speaker says—the actual words and content that come out of the speaker's mouth. Vocal is indicative of how it is said, and includes pitch, enthusiasm, inflection, and intonation. Visual is what the audience sees—the speaker's appearance, eye contact, posture, facial expressions, and gestures. All three factors go into the evaluation of speakers, but what is the relative importance of each? The results of a 1964 UCLA study by Dr. Albert Mehrabian (who has bachelors and masters degrees in engineering and a Ph.D. in psychology) indicates that the impact of the three elements is 7% verbal, 38% vocal, and 55% visual. That seems disappointing because we would like to think that the content of the presentation is most important. Realize that the numbers come from a simplified study and it is likely that that the percentages would be different for a highly technical audience. The point is that content is important, but the other elements can't be ignored. If the visual and vocal aspects of the presentation are poor, it will be perceived negatively and make it difficult for the audience to accept and pay attention to the information presented.

Here is another consideration to think about the next time you make a presentation or meet somebody. In the first seven seconds of meeting someone, people typically form a great number of subconscious opinions about the person they meet [Bai81]. This includes the person's income level, education level, competence, character, trustworthiness, personality, confidence, intelligence, work ethic, and dependability. Based on what factors are the opinions made? They include appearance, dress, posture, and speech patterns.

12.2 Preparing the Presentation

In order to make an effective presentation, the presenter must understand the subject matter (substance counts), understand the needs of the audience, and prepare the presentation. The remainder of this section provides guidance for preparing the presentation.

Analyze the Audience

An oral presentation is for the benefit of the audience, not the presenter. It is necessary to analyze and understand the audience's needs and prepare the presentation to meet them. For example, a presentation for engineering professors would likely be different from one for your family and friends. Analyzing the audience is no different than the process that one goes through when writing a document. Some questions to ask in this process are [Bai81]:

- What are they interested in?
- What do they want from your talk?
- What does the audience already know about my subject?
- What don't they know?
- What is the attitude of the audience towards me and my subject?
- What are the values of the audience?
- What do you want them to know or learn?

Understanding the needs of the audience and putting their interests first establishes credibility so that they are more willing to accept the content of the presentation.

Before creating the presentation, identify the main points that the audience should take away from it. A rule of thumb is to identify three main points for a talk, as people tend to forget more than that. Although it is not a strict rule, keep the number of points in that range, say two to five. Once the points are identified, organize the presentation to support them.

Organize the Presentation

Just like a story, a presentation has an introduction, a body, and a conclusion. This is encapsulated in the often-heard wisdom for presentations to "tell them what you are going to tell them" (the introduction), "tell it to them" (the body), and "tell them what you just told them" (the conclusion).

The introduction is absolutely critical—if the audience does not understand the presentation from the outset, they will tune out. Einstein's advice that nothing should be explained in such a way that it cannot be understood by an intelligent 12 year old is particularly relevant here. Take time to explain the problem in simple terms. Part of an effective introduction is obtaining the interest of the audience. There are many ways to accomplish this, and examples include the use of rhetorical questions or the narration of an experience that the audience can relate to. This should have to do in some way with the information being presented. Overall, the objective is to motivate the audience by describing what is being presented and why it is important. After giving motivation of the problem, an overview of the talk can be provided. The overview should be relevant to the problem at hand, not a generic one that can apply to virtually any presentation.

Structure the body of the presentation to support the main points. This is done by having a group of two to four related slides that support each of the main points. The first slide of the group provides some key ideas, followed by the remaining slides that go into more detail on the particular point. Don't make the talk unnecessarily technical or use a lot of jargon. This does not mean that it should not have technical content, but that judgment should be exercised in presenting the right amount of detail. The level of technical detail depends upon the education and experience of the audience. If it is necessary to use jargon or acronyms, make sure that they are defined for the audience. Consider alternative ways of explaining things. The use of analogies is particularly powerful when explaining complex and abstract material. One strategy is to increase the level of complexity as the talk proceeds. That way, much of the early material will be understandable to the majority of the audience, while the latter more complex material may only be understood by a small fraction. Everybody will then leave the presentation with some understanding of the content.

In a typical engineering classroom lecture, the professor usually goes through many steps in defining and deriving equations. Realize that the goals of a classroom lecture are very different from that of an oral presentation. When working with equations, don't derive or give too many intermediate steps, unless that is the point of the presentation. Provide assumptions, selected intermediate equations, and the important results. Audiences generally assume you have done your homework and derived the equations properly. There is a tendency to present equations, vaguely refer to them, and then move on. Equations should be presented for a reason, so talk about them and describe their significance. Every equation has its own story; it is the presenter's job to tell it. The same is also true of graphs and plots.

The conclusion provides the opportunity to summarize and emphasize the main points of the presentation. Again, tell them what you told them by reviewing the important points and conclusions. That way if somebody was lost during the presentation, they can understand the importance of the work. If there are recommendations to be made for future action, address them here. The conclusion is also an opportunity to explain the next steps for the project.

Lay Out the Slides

Below are pointers for the layout of the slide content.

- *Use a large font.* This ensures that information on the slides can easily be seen by the audience. 24 point or greater font is typically sufficient.

- *Have a goal of five-to-seven bullet points per page.* Avoid the tendency to cram as much information as possible on a page, which is often done so that the presenter does not neglect any material. Avoid this, and use five to seven bullet points to guide the discussion. Presentation software packages, such as Microsoft PowerPoint™, allow you to introduce bullet items one at a time, which help to keep the discussion on track.

- *Avoid fancy graphics and special effects that add no value.* Presentation packages allow the addition of fancy features, such as spiraling text and sound effects. They add little to the presentation and when overused distract from the content. The content and material are what matter the most, not fancy formatting and special effects. To quote Edward Tuft, professor emeritus at Yale,

Power Corrupts, PowerPoint Corrupts Absolutely.

His point is that fancy graphics and features are used far too much with PowerPoint presentation software, and that this overpowers both the content and the audience. [Tuf03].

- *Group slides together to make a major point.* Make the first slide the general one with key statements. The following ones should have more detailed information supporting the point.

- *Do not create a canned talk or speech.* That is acceptable in some fields, but not in engineering and science where a more extemporaneous style is the norm. Let the bullet points and other material on the slides serve as guides for what to say. Avoid the use of cue cards and do not just read directly from the slides for the presentation.

Meet the Time Constraints

Make sure that the presentation falls within the time constraints—the audience will be alienated if it is far too short or too long. The tendency is to exceed the time limit since there is so much information that the presenter wants to convey. You may be abruptly cut off and not be able to conclude the presentation if the time limit is exceeded. Think about this—how would you describe all that you know about electrical or computer engineering in ten minutes? It is challenging, but if you only had ten minutes you would probably give a brief overview of the major accomplishments made in the field. Accept that all of the information can't be conveyed in the given time and use it carefully to highlight the important material.

A heuristic is to take the length of the allotted time in minutes and divide it by two. That provides an estimate of the number of slides to prepare. Once the presentation is prepared, practice to see if it can be presented reasonably in the time allotted. Practice the talk in front of your teammates, boyfriend, girlfriend, mother, or pet rock. Be careful not to over-prepare to the point of sounded scripted. Practice the talk the night before the presentation and only do a brief review of the material right before the talk. Be sure to allow time for the question and answer session that usually occurs at the end of a presentation.

Prepare for the Question and Answer Session

One of the biggest fears of presenters is the dreaded question and answer session. This is where the audience gets to ask questions and possibly expose the presenter for what they don't know. For example, the questioner may ask *"Are you familiar with the work of Johnson and Smith from 1984 in which they proposed exactly the same idea as yours?"*

How do you prepare for this? You must be knowledgeable about the subject, but you don't need to have the answers to every possible question. It is good practice to rephrase questions that are asked for the benefit of you, the audience, and the questioner. Rephrasing the question ensures that you are answering the correct question (how many times have you been annoyed when a teacher answered the wrong question?) and provides time to think and formulate a response. It demonstrates to the questioner that you understand their question and are able to present it in a different format. If the questioner is hostile, make sure that you rephrase the question in a positive light. Rephrasing is also a courtesy for the other members of the audience who may have not heard or understood the question.

Most questions are made in good faith as the questioner is trying to clarify a point or learn more. Sometimes, questioning can become hostile or aggressive. If this happens, make sure not to respond in kind or put the questioner down. The presenter has the position of power and becomes a bully if they do this. Be sure to maintain eye contact with the questioner, smile, and remain relaxed in your responses. If you can't answer the question, admit it and don't try to come up with a phony answer. If the questioner is persistent, offer to discuss it in more detail with them after the presentation.

12.3 Project Application: Design Presentations

Examples of the three presentations that you may make during the design process are listed in Table 12.1. It is a guide of points to consider preparing presentations and should be adjusted to meet particular needs of the situation. The checklist in Table 12.2 is provided to aid in preparing for presentations.

Table 12.1 Guide for preparing design presentations. The chapters associated with the points are identified.

Presentation	Points to Consider in Preparing the Presentation
Project Proposal	*Introduction.* Provide an overview of the project and address the need, motivation, goals, and objectives. The audience is probably not familiar with the concept and it is important to describe the problem in simple and concise terms (Chapter 2).
	Problem Analysis. Indicate what the current state-of-the-art in the field is regarding the technology. If it is a new product concept, identify similar products that are available and what is unique about this one. If it is a research-oriented project, include the basic theory and address current status of work in this area (Chapter 2).
	Requirements Specifications. Address the engineering requirements and provide a justification for their selection. Describe the standards and constraints that apply to the problem (Chapter 3).
	Preliminary Design Options. Depending upon progress, some preliminary options for the design may have been developed and can be presented here (Chapter 4).
Design Review	*Introduction.* Provide a brief overview of the motivation for the project.
	Requirements. Recap the critical requirements that have to be met.
	The Proposed Design. Present the high-level design. Explain how it works and how the pieces fit together. Include design details of the sub-components and systems. Address how the proposed design meets the engineering requirements. Identify the alternatives investigated (Chapters 4, 5, and 6).
	Preliminary Test Results. Include test and prototype results (Chapter 7).
	Project Plan. By this point (if not earlier), a project plan should be in place, so consider presenting a summary of the plan (schedule, responsibilities, and cost) (Chapter 10).
Final Presentation	*Introduction.* Provide an overview or motivation for the project.
	The Final Design. Describe the final design implementation. A good way to organize is to provide a high-level overview of the design and describe how it operates. Then, provide detail and a description for each of the successive hardware/software subsystems (Chapters 5 and 6).
	Testing & Results. Describe/demonstrate the key tests and results that show the functionality of the design. Provide demonstrations if appropriate. Indicate how the final realization did or did not meet the requirements (Chapter 7).
	Conclusions. Summarize conclusions about the project and provide recommendations for further work. Indicate lessons learned.

Table 12.2 Checklist and self-assessment for oral presentation preparation. Score the elements as 1 = Strongly Disagree, 2 = Disagree, 3 = Neutral, 4= Agree, 5 = Strongly Agree.

Organization	Score
The background and needs of the audience were analyzed.	
The main points of the presentation are identified.	
The motivation is clear and would be understandable to an intelligent 12 year old.	
An overview of the presentation is provided. It is relevant to the presentation, not a generic one that can be used in any presentation.	
The body is organized to support the main points.	
The conclusion summarizes the main points and future work.	
Visual Aids	
The fonts and graphics are large enough to be seen by the audience.	
The equations are of the right number and level. The presenters are prepared to discuss any equations presented.	
The slides are arranged to support the main points.	
The presentation does not contain unnecessary graphics and special effects.	
Presentation Delivery	
The presentation has been rehearsed. It meets the time constraints and there is sufficient time for questions and answers.	
Voice projection is loud enough so that the audience can hear the presenters.	
All members participate in the presentation and have reasonably equal responsibilities. (If one team member always presents the introduction and another the technical material, it is a sign that not all members are participating equally on the project).	
The presenters do not rely on cue cards.	
The presenters are comfortable in front of an audience (Do they make good eye contact with the audience? Do the presenters move around the room or do they stand stiffly behind a podium?)	
All presenters are knowledgeable on the subject and prepared to answer questions.	
The presentation software was tested on the platform to make sure it works.	
The presenters are dressed properly for the occasion.	

12.4 Summary and Further Reading

During the design process and your professional career, you will need to communicate ideas effectively. One of the most common ways to do this is via an oral presentation. Visual, verbal, and vocal aspects impact the effectiveness of an oral presentation. Although the verbal aspect, or content, is important, the visual and vocal delivery aspects heavily influence the audience's perception and cannot be overlooked. In preparing the presentation, the needs and background of the audience should be taken into consideration and the main points to be conveyed identified. Creating the presentation is much like telling a story—there should be an introduction, a body, and a conclusion that are organized to support the main points. The slides should be grouped to support these points. Concepts should be explained as simply and clearly as possible. Increasing the complexity as the presentation proceeds will allow the presentation to reach all members of the audience to some extent. Practicing the presentation is especially important for novice presenters, and tips were provided for meeting the time constraints and preparing for question sessions.

There are many excellent resources and articles available regarding oral presentations. A concise and humorous article geared for new speakers in the technical fields is *Advice to Beginning Physics Speakers* by James Garland [Gar91]. The IEEE Transactions on Professional Communications journal addresses many aspects of communications including oral presentations. Mindtools (www.mindtools.com) has a section on communication skills that includes a preparation checklist on presentation, delivery, appearance, and visual aids. *A Good Speech is Worth a Thousand Words* by Bert Decker [Dec84] addresses right and left brain thinking as well as the three V's of giving a presentation. The article *How to Overcome Errors in Public Speaking* by John Baird [Bai81] addresses how to analyze the audience, the judgments that are made when meeting a person, the introduction, and conclusion. Other resources used in the preparation of this chapter include: *The Engineering Presentation—Some Ideas on How to Approach and Present It* [Ros93], *Handling a Hostile Audience—With Your Eyes* [Car89], and *How to Speak so that Facts Come Alive* [Ste89]. Many of the references are compiled in the book Writing and Speaking in the Technology Professions: A Practical Guide edited by David Beer [Bee03].

References

ABC01   ABC News Productions Nightline, "The Deep Dive," Video #N990713-51, dated July 13, 1999, Copyrighted 2001.

ABE03   *Criteria for Accrediting Engineering Programs*, ABET Engineering Accreditation Commission, 2003.

Abl91   R. Abler, "Good Product Designs Start with Good Specifications," *Medical Device Diagnostic Industry*, pp. 156-157, June 1991.

Ada01   J.L. Adams, Conceptual Blockbusting, 4th Edition, Perseus Publishing, 2001.

Alt99   G. Altshuller, The Innovation Algorithm: TRIZ, Systematic Innovation, and Technical Creativity, Technical Innovation Center, 1999.

Ang00   R.B. Angus, N.R. Gunderson, and T.P. Cullinane, Planning, Performing and Controlling Projects: Principles and Applications, 2nd Edition, Prentice Hall, 2000.

App01   D. Apple and K. Krumsieg, Pacific Crest Teaching Institute Handbook, 2001.

Bai81   J.E. Baird Jr., "How to Overcome Errors in Public Speaking," *IEEE Transactions on Professional Communication*, vol. PC-24, no. 2, pp. 94–98, 1981.

Bee03   D.F. Beer, editor, Writing and Speaking in the Technology Professions: A Practical Guide, Wiley-IEEE Press, 2003.

Bel94   L. Bellamy, D. Evans, D. Linder, B. McNeill, and G. Raupp, *Team Training Workbook*, Arizona State University, 1994. Available at <http://www.eas.asu.edu/~asufc/teaminginfo/teamwkbk.pdf>.

Ben01   S. Bennett and J. Skelton, Schaum's Outline of UML, McGraw Hill, 2001.

Bet04   J. Betts, R. Grayson, S. Haug, and W.. Wykoff, "The Vision Extended Car," *Penn State Erie Electrical and Computer Engineering Senior Design Report*, 2004.

Bur03   E.M. Burke and B.M. Coyner, Java Extreme Programming Cookbook, O'Reilly, 2003.

Cag02   J. Cagan and C.M. Vogel, Creating Breakthrough Products: Innovation from Product Planning to Program Approval, Financial Times/Prentice Hall, 2002.

Car89   G. Carle, "Handling a Hostile Audience–With Your Eyes," *IEEE Transactions on Professional Communication*, vol. 32, no. 1, pp. 29–31, 1989.

Cha71   N. Chapin, Flowcharts, Auerbach, 1971.

Cha02   K. Chakraborty and P. Mazumder, <u>Fault-Tolerance and Reliability Techniques for High-Density Random-Access Memories</u>, Prentice Hall PTR, 2002.

Cha02b   T. Chambers, "A Piercian Approach to Professional Ethics Instruction," *IEEE Transactions on Professional Communication*, vol. 45, no. 1, pp. 45–49, March, 2002.

Cic05   J. Ciccozzi, D. Lamb, and D. Vilairoj, "The RobotTender," *Penn State Erie Electrical and Computer Engineering Senior Design Report*, 2005.

Cro00   N. Cross, <u>Engineering Design Methods: Strategies for Product Design</u>, 3rd Edition, John Wiley & Sons, 2000.

Deb67   E. DeBono, <u>New Think: The Use of Lateral Thinking in the Generation of New Ideas</u>, New York, Basic Books, 1967.

Deb70   E. DeBono, <u>Lateral Thinking: Creativity Step by Step</u>, Harper and Row, 1970.

Dec84   B. Decker, "A Good Speech is Worth a Thousand Words," *IEEE Transactions on Professional Communication*, vol. PC-27, no. 1, pp. 32–34, March, 1984.

Deg81   R.T. DeGeorge, "Ethical Responsibilities of Engineers in Large Organizations: The Pinto Case," *Business and Professional Ethics Journal*, vol. 1, no. 1, pp. 1–14, 1981.

Del71   A.L. Delbecq and A. H. VandeVen, "A Group Process Model for Problem Identification and Program Planning," *Journal of Applied Behavioral Science VII*, pp. 466-491, July/August, 1971.

Die00   G.E. Dieter, <u>Engineering Design</u>, 3rd Edition, McGraw-Hill, 2000.

Dom01   P.G. Dominick *et al.*, <u>Tools and Tactics of Design</u>, John Wiley & Sons, 2001.

Dym94   C. Dym, <u>Engineering Design: A Synthesis of Views</u>, Cambridge University Press, 1994.

Elm94   R. Elmasri and S.B. Navathe, <u>Fundamentals of Database Systems</u>, Cummings, 1994.

Ern97   R. Ernst, <u>Embedded System Architectures in Hardware/Software Co-Design: Principles and Practice</u>, Kluwer Academic Publishers, 1997.

Ese03   T. Esek, J. Hunt, and N. Lewis, "Intel Pro 1000XF Server Testing," *Penn State Erie Electrical and Computer Engineering Senior Design Report*, 2003.

Far02   J. Farrell, <u>Programming Logic and Design</u>, 2nd Edition, Thomson Learning, 2002.

Fis04   K.J. Fisher, M.J. Lobaugh, and D.H. Parente, "An Assessment of Desired Business Knowledge Attributes for Engineering and Technology Graduates," Submitted to the *Journal of Engineering Technology*, July 2004.

Gar91   J.C. Garland, "Advice to Beginning Physics Speakers," *Physics Today*, pp. 42–45, July, 1991.

Gob99   A. Gobold, "What has Ethics to do with me? I am an Engineer," *Engineering Management Journal*, vol. 9, no. 2, pp. 81–86, 1999.

Gra85   J. Gray. "Why do Computers Stop and What Can be Done About it?," *Tandem Technical Report 85.7*, 46 pages, 1985.

Gra02   C.F. Gray and E.W. Larson, Project Management, 2nd Edition, McGraw-Hill, 2002.

Gri93   A. Griffin and J.R. Hauser, "The Voice of the Customer," *Marketing Science*, vol. 12, no. 1, pp. 1–27, 1993.

Hau88   J.R. Hauser and D. Clausing, "The House of Quality," *Harvard Business Review*, pp. 63–73, May/June, 1988.

Hna03   E.R. Hnatek, Practical Reliability of Electronic Equipment and Products, Marcel Dekker, 2003.

Hym98   B. Hyman, The Fundamentals of Engineering Design, Prentice-Hall, 1998.

IEEE Std   IEEE Standard Glossary of Software Engineering Terminology, IEEE Standard 610.12, 1990.
610.12-
1990

IEEE Std   IEEE Standard for Software Requirements Specifications, IEEE Standard 830, 1998.
830-1998

IEEE Std   IEEE Guide for Developing System Requirements Specifications, IEEE Standard 1233, 1998.
1233-1998

Jal97   P. Jalote, An Integrated Approach to Software Engineering, 2nd Edition, Spring-Verlag, New York, 1997.

Joh02   D.W. Johnson and F. Johnson, Joining Together: Group Theory and Group Skills, 8th Edition, Prentice Hall, 2002.

Kat93   J.R. Katzenbach and D.K. Smith, The Wisdom of Teams: Creating the High-Performance Organization, HarperCollins, 1993.

Kei84   D. Keirsey and M. Bates, Please Understand Me: Character and Temperament Types, Prometheus Nemesis, 1984.

Kel01   T. Kelley and J. Littman, The Art of Innovation: Lessons in Creativity from IDEO, America's Leading Design Firm, Doubleday, 2001.

Leo95   "The World-Class Engineer," Penn State University Leonhard Center, 1995.

Loc97   "The Ethics Challenge Leaders Guide," Lockheed Martin Corporation, 1997.

Mci02   "The Ten C's for Evaluating Internet Resources," University of Wisconsin-Eau Claire McIntyre Library, http://www.uwec.edu/library/Guides/tencs.html , June, 2002.

Mcl93   G.F. McClean, "Integrating Ethics and Design," *IEEE Technology and Society Magazine*, vol. 12, no. 3, pp. 19–30, 1993.

Mic91   M. Michalko, Thinkertoys, Ten Speed Press, 1991.

MIL-HDBK 217F	Military Handbook for Reliability Prediction of Equipment, United States Department of Defense, 1991.
MIL-HDBK 881	Department of Defense Handbook: Work Breakdown Structure, United States Department of Defense, 1998.
Nov00	G. Novacek, "Designing for Reliability, Maintainability, and Safety: Part 1 Getting Started," *Circuit Cellar*, Issue 125, December, 2000.
Nov01	G. Novacek, "Designing for Reliability, Maintainability, and Safety: Part 2: Digging Deeper," *Circuit Cellar*, Issue 126, January, 2001.
Osb48	A.F. Osborn, Your Creative Power: How to Use Your Imagination, Scribner, 1948.
Osb63	A.F. Osborn, Applied Imagination: Principles and Procedures of Creative Problem-Solving, Scribner, 1963.
Oul04	The Glossary of Software Vulnerability Testing, The University of Oulu, Department of Electrical and Information Engineering webpage, http://www.ee.oulu.fi/research/ouspg/sage/glossary, 2004.
Pag88	M. Page-Jones, The Practical Guide to Structured Systems Design, 2nd Edition, Yourdan Press Computing Series, 1988.
Pet97	T. Peters, The Circle of Innovation: You Can't Shrink Your Way to Greatness, Knopf, 1997.
Pet00	J.F. Peters and W. Pedrycz, Software Engineering: An Engineering Approach, John Wiley and Sons, 2000.
Pri90	S. Pritchard, "Friends, Romans, Cost Engineers, … Can We Talk," *Cost Engineering*, February, 1990.
Pug90	S. Pugh, Total Design: Integrated Methods for Successful Product Engineering, Addison-Wesley, 1990.
Ros93	R. Rosenburg, "The Engineering Presentation–Some Ideas on How to Approach and Present It," *IEEE Transactions on Professional Communication*, vol. PC-26, no. 1, pp. 191–193, December, 1993.
Roy70	W.W. Royce, "Managing the Development of Large Software Systems: Concepts and Techniques," *Proceedings of the IEEE WESTCON*, 1970.
Rum98	J. Rumbaugh, I. Jacobson, and G. Booch, Unified Modeling Language Reference Manual, Addison-Wesley, 1998.
Run99	D.L. Runnels, "How to Write Better Test Cases," International Conference on Software Testing and Review, 12 pages, 1999.
Sat88	T. Satay, Decision Making for Leaders: The Analytical Hierarchy Process for Decisions in a Complex World, University of Pittsburgh, 1988.

Sat02   J.W. Satzinger, <u>Systems Analysis and Design in a Changing World</u>, Course Technology, 2002.

Sch98   K. Schrock, "The ABCs of Website Evaluation," *Classroom Connect*, pp. 4–6, December 1998/January, 1999. Available at http://kathyschrock.net/abceval/.

Sed04   A. Sedra and K.C. Smith, <u>Microelectronic Circuits</u>, Oxford University Press, 2004.

Sha94   J. Shapiro, "Profession/Profile of George Heilmeier," *IEEE Spectrum*, pp. 56–58, June, 1994.

Slo91   P. Sloane , <u>Lateral Thinking Puzzlers</u>, Sterling, 1991.

Slo93   P. Sloane and D. MacHale, <u>Challenging Lateral Thinking Puzzles</u>, Sterling, 1993.

Slo94   P. Sloane, <u>Great Lateral Thinking Puzzles</u>, Sterling, 1994.

Smi04   K.A. Smith, <u>Teamwork and Project Management</u>, 2nd Edition, McGraw-Hill, 2004.

Som01   I. Sommerville, <u>Software Engineering</u>, 6th Edition, Addison Wesley, 2001.

Sta01   J. Stadtmiller, <u>Electronics: Project Management and Design</u>, Prentice Hall, 2001.

Ste89   M.A. Stettner, "How to Speak so Facts Come Alive," *Chemical Engineering*, pp. 195, September, 1989.

Ste99   W.P. Stevens, G.J. Myers, and L.L. Constantine, "Structured Design," *IBM Systems Journal*, vol. 13, no. 2, pp. 231–256, 1999 (Reprinted from 1974).

Str02   C.E. Stroud, <u>A Designer's Guide to built-In Self Test</u>, Kluwer Academic Publishers, 2002.

Tel96   Telcordia SR332, <u>Reliability Prediction Procedure for Electronic Equipment</u>, 1996.

Tuc65   B. Tuckman, "Developmental Sequence in Small Groups," *Psychological Bulletin*, vol. 63, pp. 384–399, 1965.

Tuf03   E. Tufte, "PowerPoint is Evil: Power Corrupts, PowerPoint Corrupts Absolutely," *Wired Magazine*, September, 2003. Available at http://www.wired.com/wired/archive/11.09/ppt2.html.

Ulr03   K.T. Ulrich and S.D. Eppinger, <u>Product Design and Development</u>, 3rd Edition, McGraw-Hill, 2003.

Van01   A. Van Gorp and I. Van De Poel, "Ethical Considerations in the Engineering Design Process," *IEEE Technology and Society Magazine*, vol. 20, no. 3, pp. 15–22, 2001.

Vol03   *Volere Requirements Specification Template*, Version 9, 2003.

Wak00   J. Wakerly, <u>Digital Design: Principles and Practice</u>, 3rd Edition, Prentice Hall, 2000.

Wil95   D. Wilemon, "Cross-Functional Teamwork in Technology Based Organizations," *IEEE Engineering Management Conference*, pp. 74–79, 1995.

Wol01   W. Wolf, <u>Computers and Components: Principles of Embedded Computing System Design</u>, Morgan Kaufmann, 2001.

Wol02   W. Wolf, <u>Modern VLSI Design</u>, Prentice Hall, 2002.

Yal01   S. Yalamanchili, <u>Introductory VHDL: From Simulation to Synthesis</u>, Prentice Hall, 2001.

You89   E. Yourdan, <u>Modern Structured Analysis</u>, Yourdan Press Computing Series, Prentice Hall, 1989.

Appendix A Glossary

acceptance test
An acceptance test verifies that the system meets the *Requirements Specification* and stipulates the conditions under which the customer will accept the system (Chapter 7).

activity on node
A form of a *network diagram* used in a project plan. In the Activity on Node (AON) form, activities are represented by nodes and the dependencies by arrows (Chapter 10).

activity
An activity is a combination of a *task* and its associated *deliverables* that is part of a project plan (Chapter 10).

activity view
The activity view is part of the *Unified Modeling Language*. It is characterized by an activity diagram; its *intention* is to describe the sequencing of processes required to complete a task (Chapter 6).

artifact
System, component, or process that is the end-result of a design (Chapter 2).

automated script test
An automated script test is a sequence of commands given to a unit under test. For example, a test may consist of a sequence of inputs that are provided to the unit, where the outputs for each input are then verified against pre-specified values (Chapter 7).

baseline requirements
The original set of requirements that are developed for a system (Chapter 3).

black box test
A test that is performed without any knowledge of internal workings of the unit under test (Chapter 7).

bottom-up design
An approach to system design where the designer starts with basic components and synthesizes them to achieve the design objectives. This is contrasted to *top-down* design (Chapter 5).

Bohrbug
Bohrbugs are reliable *bugs*, in which the error is always in the same place. This is analogous to the electrons in the Bohr atomic model which assume a definite orbit (Chapter 7).

brainstorming
A freeform approach to concept generation that is often done in groups. This process employs five basic rules: 1) no criticism of ideas, 2) wild ideas are encouraged, 3) quantity is stressed over quality, 4) build upon the ideas of others, and 5) all ideas are recorded (Chapter 4).

break-even point
The break-even point is the point where the number of units sold is such that there is no profit or loss. It is determined from the total costs and revenue (Chapter 10).

bug	A problem or error in a system that causes it to operate incorrectly (Chapter 7).
cardinality ratio	The cardinality ratio describes the multiplicity of the entities in a relationship. It is applied to *entity relationship diagrams* and Unified Modeling Language *static view diagrams* (Chapter 6).
class	Classes are used in object-oriented system design. A class defines the attributes and methods (functions) of an *object* (Chapter 6).
cohesion	Refers to how focused a module is—highly cohesive systems do one or a few things very well. Also see *coupling* (Chapter 5).
component design specification	See *subsystem design specification* (Chapter 3).
concept fan	A graphical tree representation of design decisions and potential solutions to a problem. Also see *concept table* (Chapters 1 and 4).
concept generation	A phase in the *design process* where many potential solutions to solve the problem are identified (Chapter 1).
concept table	A tool for generating concepts to solve a problem. It allows systematic examination of different combinations, arrangements, and substitutions of different elements for a system. Also see *concept fan* (Chapter 4).
conditional rule-based ethics	An ethics system in which there are certain conditions under which an individual can break a rule. This is generally because it is believed that the moral good of the situation outweighs the rule. Also see *rule-based ethics* (Chapter 11).
constraint	A restriction that impacts the design but is not a specific function that the system must deliver (sometimes referred to as *non-functional requirements*). Constraints are broader in scope than *engineering requirements*, including social, ethical, and political considerations that may not be *verifiable* (Chapter 3).
controllability	A principle that applies to testing. Controllability is the ability to set any node of the system to a prescribed value (Chapter 7).
copyright	Copyrights protect published works such as books, articles, music, and software. A copyright means that others cannot distribute copyrighted material without permission of the owner (Chapter 11).
coupling	Modules are coupled if they depend upon each other in some way to operate properly. Coupling is the extent to which modules or subsystems are connected. See also *cohesion* (Chapter 5).
creative design	A formal categorization of design projects. Creative designs represent new and innovative designs (Chapter 2).
critical path	The path with the longest duration in a project plan. It represents the minimum time required to complete the project (Chapter 10).

cross-functional team	Cross-functional teams are those that are composed of people from different organizational functions, such as engineering, marketing, and manufacturing. Also see *multi-disciplinary team* (Chapter 9).
data dictionary	A dictionary of data contained in a *data flow diagram*. It contains specific information on the data flows and is defined using a formal language (Chapter 6).
data flow diagram	The *intention* of a data flow diagram (DFD) is to model the processing and flow of data inside a system (Chapter 6).
decision matrix	A matrix that is used to evaluate and rank concepts. It integrates both the user-needs and the technical merits of different concepts (Chapter 4).
derating	A decrease in the maximum amount of power that can be dissipated by a device. The amount of derating is based upon operating conditions, notably increases in temperature (Chapter 8).
deliverable	Deliverables are entities that are delivered to the project based upon completion of *tasks*. Also see *activity* (Chapter 10).
descriptive design process	Describes typical activities involved in realizing designs with less emphasis on exact sequencing than a *prescriptive design process* (Chapter 1).
design architecture	The main (Level 1) organization and interconnection of modules in a system (Chapter 5).
design phase	Phase in the *design process* where the technical solution is developed, ultimately producing a detailed system design. Upon its completion, all major systems and subsystems are identified and described using an appropriate model (Chapter 1).
design process	The steps required to take an idea from concept to realization of the final system. It is a problem-solving methodology that aims to develop a system that best meets the customer's need within given constraints (Chapter 1).
design space	The space, or collection, of all possible solutions to a design problem (Chapter 2).
detailed design	A phase in the technical design where the problem can be decomposed no further and the identification of elements such as circuit components, logic gates, or software code takes place (Chapter 5).
engineering requirement	A requirement of the system that applies to the technical aspects of the design. An engineering requirement should be abstract, unambiguous, verifiable, and traceable (Chapter 3).
entity relationship diagram (ERD)	An ERD is used to model database systems. The *intention* of an ERD is to catalog a set of related objects (entities), their attributes, and the relationships between them (Chapter 6).

entity relationship matrix	A matrix that is used to identify relationships between entities in a database system (Chapter 6).
ethics	Philosophy that studies *morality*, the nature of good and bad, and choices to be made (Chapter 11).
event	An event is an occurrence at a specific time and place that needs to be remembered and taken into consideration in the system design (Chapter 6).
event table	A table that is used to store information about *events* in the system. It includes information regarding the event trigger, the source of the event, and process triggered by the event (Chapter 6).
failure function	The failure function, $F(t)$, is a mathematical function that provides the probability that a device has failed at time t (Chapter 8).
failure rate	The failure rate, $\lambda(t)$, for a device is the expected number of device failures that will occur per unit time (Chapter 8).
fixed costs	Fixed costs are those that are constant regardless of the number of units produced and cannot be directly charged to a process or activity (Chapter 10).
float	The amount of *slippage* that an activity in a project plan can experience without it becoming part of a new *critical path* (Chapter 10).
flowchart	A modeling diagram whose intention is to visually describe a process or algorithm, including its steps and control (Chapter 6).
functional decomposition	A design technique in which a system is designed by determining its overall functionality and then iteratively decomposing it into component subsystems, each with its own functionality (Chapter 5).
functional requirement	A *subsystem design specification* that describes the inputs, outputs, and functionality of a system or component (Chapters 3 and 5).
Gantt chart	Gantt charts are a bar graph representation of a project plan where the activities are shown on a timeline (Chapter 10).
Heisenbugs	Heisenbugs are *bugs* that are not always reproducible with the same input. This is analogous to the Heisenberg Uncertainty Principle, in which the position of an electron is uncertain (Chapter 7).
high-performance team	A team that significantly outperforms all similar teams. Part of the Katzenbach and Smith team model (Chapter 9).
integration test	An integration test is performed after the units of a system have been constructed and tested. The integration test verifies the operation of the integrated system behavior (Chapter 7).
intention	The intention of a model is the target behavior that it aims to describe (Chapter 6).

interaction view	The interaction view is part of the *Unified Modeling Language*. Its *intention* is to show the interaction between objects. It is characterized by collaboration and sequence diagrams (Chapter 6).
key attribute	An attribute for an entity in a database system that uniquely identifies an instance of the entity (Chapter 6).
lateral thinking	A thought process that attempts to identify creative solutions to a problem. It is not concerned with developing the solution for the problem, or right or wrong solutions. It encourages jumping around between ideas. It is contrasted to *vertical thinking* (Chapter 4).
liable	Required to pay monetary damages according to law (Chapter 11).
marketing requirement (specifications)	A statement that describe the needs of the customer or end-user of a system. They are typically stated in language that the customer would use (Chapters 2 and 3).
maintenance phase	Phase in the *design process* where the system is maintained, upgraded to add new functionality, or design problems are corrected (Chapter 1).
matrix test	A matrix test is a test that is suited to cases where the inputs submitted are structurally the same and differ only in their values (Chapter 7).
mean time to failure	The mean time to failure (MTTF) is a mathematical quantity which answers the question, *"On average how long does it take for a device to fail?"* (Chapter 8).
module	A block, or subsystem, in a design that performs a function (Chapter 5).
morals	The *principles* of right and wrong and the decisions that derive from those principles (Chapter 11).
multi-disciplinary team	In general, a multi-disciplinary team is one in which the members have complementary skills and the team may have representation from multiple technical disciplines. Also see *cross-functional team* (Chapter 9).
negligence	Failure to exercise caution, which in the case of design could be in not following reasonable standards and rules that apply to the situation (Chapter 11).
network diagram	A network diagram is a directed graph representation of the activities and dependencies between them for a project (Chapter 10).
Nominal Group Technique (NGT)	A formal approach to brainstorming and meeting facilitation. In NGT, each team member silently generates ideas that are reported out in a round-robin fashion so that all members have an opportunity to present their ideas. Concepts are selected by a multi-voting scheme with each member casting a predetermined number of votes for the ideas. The ideas are then ranked and discussed (Chapters 4 and 9).

non-disclosure agreement	An agreement that prevents the signer from disseminating information about a company's products, services, and trade secrets (Chapter 11).
object	Objects represent both data (attributes) and the methods (functions) that can act upon data. An object represents a particular instance of a *class*, which defines the attributes and methods (Chapter 6).
object type	Characteristic of a model used in design. The object type is capable of encapsulating the actual components used to construct the system (Chapter 6).
objective tree	A hierarchical tree representation of the customer's needs. The branches of the tree are organized based upon functional similarity of the needs (Chapter 2).
observability	This principle applies to testing. Observability is the ability to observe any node of a system (Chapter 7).
over-specificity	This refers to applying targets for *engineering requirements* that go beyond what is necessary for the system. Over-specificity limits the size of the *design space* (Chapter 3).
pairwise comparison	A method of systematically comparing all customer needs against each other. A comparison matrix is used for the comparison and the output is a scoring of each of the needs (Chapter 2).
parallel system	A system that contains multiple modules performing the same function where a single module would suffice. The overall system functions correctly when any one of the submodules is functioning (Chapter 8).
patent	A patent is a legal device for protecting a design or invention. If a patent is held for a technology, others cannot use it without permission of the owner (Chapter 11).
path-complete coverage	Path-complete coverage is where every possible *processing path* is tested (Chapter 7).
performance requirement	A particular type of *engineering requirement* that specifies performance related measures (Chapter 3).
physical view	The physical view is part of the *Unified Modeling Language*. Its *intention* is to demonstrate the physical components of a system and how the logical views map to them. It is characterized by a component and deployment diagram (Chapter 6).
potential team	A team where the sum effort of the team equals that of the individuals working in isolation. Part of the Katzenbach and Smith team model (Chapter 9).
prescriptive design process	An exact process, or systematic recipe, for realizing a system. Prescriptive design processes are often algorithmic in nature and expressed using flowcharts with decision logic (Chapter 1).

principle

Fundamental rules or beliefs that govern behavior, such as the Golden Rule (Chapter 11).

problem identification

The first phase in the design process where the problem is identified, the customer needs identified, and the project feasibility determined (Chapter 1).

processing path

A processing path is a sequence of consecutive instructions or states encountered while performing a computation. They are used to develop test cases (Chapter 7).

prototyping and construction phase

Phase in the *design process* in which different elements of the system are constructed and tested. The objective is to model some aspect of the system, demonstrating functionality to be employed in the final realization (Chapter 1).

pseudo-team

An under-performing team where the sum effort of the team is below that of the individuals working in isolation. Part of the Katzenbach and Smith team model (Chapter 9).

real team

A team where the sum effort of the team exceeds that of the individuals working in isolation. Part of the Katzenbach and Smith team model (Chapter 9).

redundancy

A design has redundancy if it contains multiple modules performing the same function where a single module would suffice. Redundancy is used to increase *reliability* (Chapter 8).

reliability

Reliability, $R(t)$, is the probability that a device is functioning properly (has not failed) at time t (Chapter 8).

research phase

Phase in the *design process* where research on the basic engineering and scientific principles, related technologies, and existing solutions for the problem are explored (Chapter 1).

Requirements Specification

A collection of engineering and marketing requirements that a system must satisfy in order for it to meet the needs of the customer or end-user. Alternate terms that are used for the Requirements Specification are the *Product Design Specification* and the *Systems Requirements Specification* (Chapter 1 and 3).

reverse-engineering

Process where a device or process is taken apart to understand how it works (Chapter 11).

routine design

A formal categorization of design projects. They represent the design of artifacts for which theory and practice are well-developed (Chapter 2).

rule-based ethics

Rule-based ethics are based upon a set of rules that can be applied to make decisions. In the strictest form, they are considered to be absolute in terms of governing behavior (Chapter 11).

satisfice

Satisfice means that a solution may meet the design requirements, but not be the optimal solution (Chapter 11).

series system	A system in which the failure of a single component (or subsystem) leads to failure of the overall system (Chapter 8).
situational ethics	Situational ethics are where decisions are made based on whether they produce the highest good for the person (Chapter 11).
slippage	Refers to an activity in a project plan taking longer than its planned time to complete. See also *critical path* and *float* (Chapter 10).
standards	A standard or established way of doing things. Standards ensure that products work together, from home plumbing fixtures to the modules in a modern computer. They ensure the health and safety of products (Chapter 3).
state	The state of a system represents the net effect of all the previous inputs to the system. Since the state characterizes the history of previous inputs, it is often synonymous with the word memory (Chapter 6).
state diagram (machine)	Diagram used to describe systems with memory. It consists of states and transitions between states (Chapter 6).
static view	The static view is part of the **Unified Modeling Language**. The *intention* of the static view is to show the classes in a system and their relationships. The static view is characterized by a class diagram (Chapter 6).
step-by-step test	A step-by-step test case is a prescription for generating a test and checking the results. It is most effective when the test consists of a complex sequence of steps (Chapter 7).
strengths and weakness analysis	A technique for the evaluation of potential solutions to a design problem where the strengths and weaknesses are identified (Chapter 4).
structure charts	Specialized block diagrams for visualizing functional software designs. They employ input, output, transform, coordinate, and composite modules (Chapter 5).
strict liability	A form of *liability* that focuses only on the product itself—if the product contains a defect that caused harm, the manufacturer is liable (Chapter 11).
stub	A stub is a device that is used to simulate a subcomponent of a system during testing. Stubs simulate inputs or monitor outputs from the unit under test (Chapter 7).
subsystem design specification	Engineering requirements for subsystems that are constituents of a larger, more complex system (Chapter 3).
system integration	Phase in the *design process* where all of the subsystems are brought together to produce a complete working system (Chapter 1).
task	Tasks are actions that accomplish a job as part of a project plan. Also see *activity* and *deliverable* (Chapter 10).

Team Process Guidelines	Guidelines developed by a team that govern their behavior and identify expectations for performance (Chapter 9).
technical specification	A list of the technical details for a given system, such as operating voltages, processor architecture, and types of memory. The technical specification is fundamentally different from a requirement in that it indicates what was achieved in the end versus what a system needs to achieve from the outset. (Chapter 3).
test coverage	Test coverage is the extent to which the test cases cover all possible *processing paths* (Chapter 7).
test phase	Phase in the design process where the system is tested to demonstrate that it meets the requirements (Chapters 1 and 7).
testable	A design is testable when a failure of a component or subsystem can be quickly located. A testable design is easier to debug, manufacture, and service in the field (Chapter 7).
top-down design	An approach to design in which the designer has an overall vision of what the final system must do, and the problem is partitioned into components, or subsystems that work together to achieve the overall goal. Then each subsystem is successively refined and partitioned as necessary. This is contrasted to *bottom-up* design (Chapter 5).
tort	The basis for which a lawsuit is brought forth (Chapter 11).
trade secret	An approach to protecting intellectual property where the information is held secretly, without *patent* protection, so that a competitor cannot access it (Chapter 11).
under-specificity	This refers to a state of the *Requirements Specification*. When it is under-specified, requirements do not meet the needs of the user and/or embody all of the requirements needed to implement the system (Chapter 3).
Unified Modeling Language (UML)	A modeling language that captures the best practices of object-oriented system design. It encompasses six different system views that can be used to model electrical and computer systems (Chapter 6).
unit test	A unit test is a test of the functionality of a system module in isolation. It establishes that a subsystem performs a single unit of functionality to some specification (Chapter 7).
use-case view	The use-case view is part of the *Unified Modeling Language*. Its *intention* is to capture the overall behavior of the system from the user's point of view and to describe cases in which the system will be used (Chapter 6).
utilitarian ethics	In utilitarian ethics, decisions are made based upon the decision that brings about the highest good for all, relative to all other decisions (Chapter 11).

validation	The process of determining whether the requirements meet the needs of the user (Chapter 3).
value	A value is something that a person or group believes to be valuable or worthwhile. Also see *principles* and *morals* (Chapter 11).
variable costs	Variable costs vary depending upon the process or items being produced, and fluctuate directly with the number of units produced (Chapter 10).
variant design	A formal categorization of design projects. They represent the design of existing systems, where the intent is to improve performance or add features (Chapter 2).
verifiable	Refers to a property of an engineering requirement. It means that there should be a way to measure or demonstrate that the requirement is met in the final system realization (Chapter 3).
vertical thinking	A linear, or sequential, thought process that proceeds logically towards the solution of a problem. It seeks to eliminate incorrect solutions. It is contrasted to *lateral thinking* (Chapter 4).
whistleblower	A person who goes outside of their company or organization to report an ethical or safety problem (Chapter 11).
white box test	White box tests are those that are conducted with knowledge of the internal working of the unit under test (Chapter 7).
work breakdown structure	The work breakdown structure (WBS) is a hierarchical breakdown of the tasks and deliverables that need to be completed in order to accomplish a project (Chapter 10).
working group	A group of individuals working in isolation, who come together occasionally to share information. Part of the Katzenbach and Smith team model (Chapter 9).

Appendix B Component Failure Rate Data

This appendix contains failure rate data for selected electronic components in support of Chapter 8. The information was abstracted from the <u>Military Handbook for Reliability Prediction</u> [MIL-HDBK-217F] and contains information for devices that are relevant to this book. The information is a close facsimile to MIL-HDBK-217F, but in some instances comments are added to help interpret the information. Furthermore, MIL-HDBK-217F tends to present both empirical estimation formulas and tabular data computed from the formulas. The tabular data is generally not included here for simplicity of presentation, and is readily obtained from the formulas supplied. Data is presented for the following devices:

- Analog components: resistors and capacitors.
- Discrete semiconductors: diodes, bipolar transistors, and field effect transistors.
- Microcircuits: gate/logic arrays and microprocessors.

B.1 Environmental Use

All of the devices presented in this appendix have an environmental factor that is used to estimate the failure rate. The environmental factors are based on the 14 categories in Table B.1.

Table B.1 Environmental symbols and descriptions taken directly from MIL-HDBK-217F.

Environment	Description
G_B – Ground, Benign	Nonmobile, temperature and humidity controlled environments readily accessible to maintenance; includes laboratory instruments and test equipment, medical electronic equipment, business and scientific computer complexes, and missile and support equipment in ground silos.
G_F – Ground, Fixed	Moderately controlled environments such as installation in permanent racks with adequate cooling air and possible installation in unheated buildings; includes permanent installation of air traffic control radar and communication facilities.
G_M – Ground, Mobile	Equipment installed in wheeled or tracked vehicles and equipment manually transported; includes tactical missile ground support equipment, mobile communication equipment, tactical fire direction systems, handheld communications equipment, laser designations and range finders.

Table B.1 Environmental symbols and descriptions taken directly from MIL-HDBK-217F, cont'd.

Environment	Description
N_S – Naval, Sheltered	Includes sheltered or below deck conditions on surface ships and equipment installed in submarines.
N_U – Naval, Unsheltered	Unprotected surface shipborne equipment exposed to weather conditions and equipment immersed in salt water. Includes sonar equipment and equipment installed on hydrofoil vessels.
A_{IC} – Airborne, Inhabited Cargo	Typical conditions in cargo compartments that can be occupied by an aircrew. Environment extremes of pressure, temperature, shock, and vibration are minimal. Examples include long mission aircraft such as the C130, C5, B52, and C141. This category also applies to inhabited areas in lower performance smaller aircraft such as the T38.
A_{IF} – Airborne, Inhabited Fighter	Same as A_{IC} but installed on high performance aircraft such as fighters and interceptors. Examples include the F15, F16, F111, F/A 18 and A10 aircraft.
A_{UC} - Airborne, Uninhabited Cargo	Environmentally uncontrolled areas that cannot be inhabited by crew during flight. Environmental extremes of pressure, temperature, and shock may be severe. Examples include uninhabited areas of long aircraft such as the C130, C5, B52, and C141. This category also applies to uninhabited areas in lower performance smaller aircraft such as the T38.
A_{UF} - Airborne, Uninhabited Fighter	Same as A_{UC} but installed on high performance aircraft such as fighters and interceptors. Examples include the F15, F16, F111, and A10 aircraft.
A_{RW} – Airborne, Rotary Winged	Equipment installed on helicopters. Applies to both internally and externally mounted equipment such as laser designators, fire control systems, and communications equipment.
S_F – Space, Flight	Earth orbital. Approaches benign ground conditions. Vehicle neither under powered flight nor in atmospheric re-entry; includes satellites and shuttles.
M_F – Missile, Flight	Conditions related to powered flight of air breathing missiles, cruise missiles, and missiles in unpowered free flight.
M_L – Missile, Launch	Severe conditions related to missile launch (air, ground, and sea), space vehicle boost into orbit, and vehicle re-entry and landing by parachute. Also applies to solid rocket motor propulsion powered flight, and torpedo and missile launch from submarines.
C_L – Cannon, Launch	Extremely severe conditions related to cannon launching of 155 mm and 5 inch guided projectiles. Conditions apply to the projectile from launch to target impact.

B.2 Analog Components: Resistors and Capacitors

This section contains failure rate data for resistors and capacitors.

B.2.1 Resistors: Fixed Composition, Fixed Film, and Wirewound

The failure rate is given by the following relationship

$$\lambda = \lambda_b \pi_R \pi_Q \pi_E \frac{\text{failures}}{10^6 \text{ hours}}.$$

λ_b - Base Failure Rate

Resistor Type	λ_b $\left(T = \text{ambient temperature in } °C, S = \dfrac{\text{operating power}}{\text{device power rating}}\right)$
Fixed Composition	$\lambda_b = 4.5 \times 10^{-9} \exp\left(12\left(\dfrac{T+273}{343}\right)\right) \exp\left(\dfrac{S}{0.6}\left(\dfrac{T+273}{273}\right)\right)$
Fixed Film	$\lambda_b = 3.25 \times 10^{-4} \exp\left(\left(\dfrac{T+273}{343}\right)^3\right) \exp\left(S\left(\dfrac{T+273}{273}\right)\right)$
Wirewound	$\lambda_b = 0.0031 \exp\left(\left(\dfrac{T+273}{398}\right)^{10}\right) \exp\left(\left(S\left(\dfrac{T+273}{273}\right)\right)^{1.5}\right)$

Resistance Factor - π_R

Fixed Composition or Fixed Film Resistors		Wirewound Resistors	
Resistance Range	π_R	Resistance Range	π_R
$\leq 100\text{k}\Omega$	1.0	$\leq 10\text{k}\Omega$	1.0
$100\text{k}\Omega$ to $\leq 1\text{M}\Omega$	1.1	$10\text{k}\Omega$ to $\leq 100\text{k}\Omega$	1.7
$1\text{M}\Omega$ to $\leq 10\text{M}\Omega$	1.6	$100\text{k}\Omega$ to $\leq 1\text{M}\Omega$	3.0
$> 10\text{M}\Omega$	2.5	$> 1\text{M}\Omega$	5.0

Quality Factor - π_Q

Quality	π_Q
S	0.03
R	0.1
P	0.3
M	1.0
MIL-R	5.0
Lower	15

Environmental factor - π_E

Environment	π_E - Fixed Composition	π_E - Fixed Film	π_E - Wirewound
G_B	1	1	1
G_F	3	2	2
G_M	8	8	11
N_S	5	4	5
N_U	13	14	18
A_{IC}	4	4	15
A_{IF}	5	8	18
A_{UC}	7	10	28
A_{UF}	11	18	35
A_{RW}	19	19	27
S_F	0.50	0.20	0.80
M_F	11	10	14
M_L	27	28	38
C_L	490	510	610

B.2.2 Capacitors: Fixed, Ceramic, and General Purpose

The failure rate is given by the following relationship

$$\lambda = \lambda_b \pi_{CV} \pi_Q \pi_E \frac{\text{failures}}{10^6 \text{ hours}} .$$

For all factors T = ambient temperature (°C) and $S = \dfrac{\text{operating voltage}}{\text{rated voltage}}$, where the operating

voltage is the sum of the DC and peak AC voltage.

Base Failure Rate - λ_b

$$\lambda_b = 0.0003\left[\left(\frac{S}{0.3}\right)^3 + 1\right]\exp\left(\frac{T+273}{358}\right), \text{ for } T = 85°C \text{ max rated}$$

$$\lambda_b = 0.0003\left[\left(\frac{S}{0.3}\right)^3 + 1\right]\exp\left(\frac{T+273}{398}\right), \text{ for } T = 125°C \text{ max rated}$$

$$\lambda_b = 0.0003\left[\left(\frac{S}{0.3}\right)^3 + 1\right]\exp\left(\frac{T+273}{423}\right), \text{ for } T = 150°C \text{ max rated}$$

Capacitance Factor - π_{CV}

$$\pi_{CV} = 0.41 \times C^{0.11}$$

C = capacitance

Quality Factor - π_Q

Quality	π_Q
S	0.030
R	0.10
P	0.30
M	1.0
L	3.0
MIL	3.0
Lower	10

Environmental Factor - π_E

Environment	π_E
GB	1
GF	2
GM	9
NS	5
NU	15
AIC	4
AIF	4
AUC	8
AUF	12
ARW	20
SF	0.40
MF	13
ML	34
CL	610

B.3 Microelectronic Devices

For all microelectronic devices it is necessary to compute the junction temperature (T_J) of the silicon in order to determine the temperature factor. The junction temperature is determined as follows:

$$T_J = T_A + \theta_{JA} P_D$$

T_A = ambient temperature

θ_{jA} = junction to ambient thermal resistance (obtained from manufacturer datasheet)

P_D = power dissipated in the device

Author's Note: The equations above are slightly different than those found in MIL-HDBK-217F. The junction to ambient thermal resistance is used here, instead of junction to case as in the original. In addition, the ambient temperature is used in place of the case temperature. This is more general and is consistent with the presentation in Chapter 8. The junction to case resistance could be used, along with the case temperature. See Section 8.2 for more detailed coverage of this thermal model.

The part quality descriptors in Table B.1 are used to find the quality factors.

Table B.1 Part quality descriptors for microelectronic devices.

JANTXV	Full device testing as specified by the MIL-S-19500 specification, including Screening and Groups A, B, and C.
JANTX	Identical to JANTXV, except does not include the 100% precap visual inspection contained in Screening.
JAN	Testing as defined by MIL-S-19500, including Groups A, B, and C, but not including Screening.
Lower	All hermetically packaged devices.
Plastic	All devices encapsulated with organic materials.

B.3.1 Diodes: Low Frequency

The failure rate is given by the following relationship

$$\lambda = \lambda_b \pi_T \pi_s \pi_C \pi_Q \pi_E \ \frac{\text{failures}}{10^6 \ \text{hours}} \ .$$

Base Failure Rate - λ_b

Diode Type/Application	λ_b
General purpose analog	0.0038
Switching	0.0010
Power rectifier, fast recovery	0.069
Power rectifier, Schottky power diode	0.0030
Power rectifier with high voltage stacks	0.0050/Junction
Transient suppressor/varistor	0.0013
Current regulator	0.0034
Voltage regulator and voltage reference (avalanche and zener)	0.0020

Temperature Factor - π_T

For general purpose analog, switching, fast recovery, power rectifier, and transient suppressor applications

$$\pi_T = \exp\left(-3091\left(\frac{1}{T_J + 273} - \frac{1}{298}\right)\right).$$

For voltage regulator, voltage reference, and current regulator applications

$$\pi_T = \exp\left(-1925\left(\frac{1}{T_J + 273} - \frac{1}{298}\right)\right).$$

Electrical Stress Factor - π_S

$\pi_{s'} = 0.054, \quad V_S \leq 0.3$

$\pi_{s'} = V_S^{2.43}, \quad 0.3 < V_s \leq 1$

V_s = voltage stress ratio = $\dfrac{\text{applied voltage}}{\text{rated voltage}}$

voltage is the diode reverse voltage

Contact Construction Factor - π_C

Contact Construction	π_C
Metallurgically bonded	1.0
Non-metallurgically bonded and spring loaded contacts	2.0

Quality Factor - π_Q

Quality	π_Q
JANTXV	0.7
JANTX	1.0
JAN	2.4
Lower	5.5
Plastic	8.0

Environmental Factor - π_E

Environment	π_E
GB	1
GF	6
GM	9
NS	9
NU	19
AIC	13
AIF	29
AUC	20
AUF	43
ARW	24
SF	0.50
MF	14
ML	32
CL	320

B.3.2 Diodes: High Frequency (microwave, RF)

The failure rate is given by the following relationship

$$\lambda = \lambda_b \pi_T \pi_A \pi_R \pi_Q \pi_E \; \frac{\text{failures}}{10^6 \text{ hours}}.$$

Base Failure Rate - λ_b

Diode Type	λ_b
Si IMPATT ($\leq$ 35 GHz)	0.22
Gunn/Bulk Effect	0.18
Tunnel and Back	0.0023
PIN	0.0081
Schottky Barrier	0.027
Varactor and Step Recovery	0.0025

Application Factor - π_A

Application	π_A
Varactor, voltage control	0.50
Varactor, multiplier	2.5
All other diodes	1.0

Quality Factor - π_Q

Quality	π_Q Not Shottky	π_Q Shottky
JANTXV	0.5	0.5
JANTX	1.0	1.0
JAN	5.0	1.8
Lower	25.0	2.5
Plastic	50.0	-

Temperature Factor - π_T

All types except IMPATT

$$\pi_T = \exp\!\left(-2100\!\left(\frac{1}{T_J + 273} - \frac{1}{298}\right)\right),$$

and for IMPATT

$$\pi_T = \exp\!\left(-5260\!\left(\frac{1}{T_J + 273} - \frac{1}{298}\right)\right).$$

Power Rating Factor - π_R

$\pi_R = 0.326\ln(P_R) - 0.25$, PIN diodes

$\pi_R = 1.0$, all other diodes

Environmental Factor - π_E

Environment	π_E
G$_B$	1
G$_F$	2
G$_M$	5
N$_S$	4
N$_U$	11
A$_{IC}$	4
A$_{IF}$	5
A$_{UC}$	7
A$_{UF}$	12
A$_{RW}$	16
S$_F$	0.5
M$_F$	9
M$_L$	24
C$_L$	250

B.3.3 Transistors: Bipolar Junction, Low Frequency (≤ 200MHz)

The failure rate is given by the following relationship

$$\lambda = \lambda_b \pi_T \pi_A \pi_R \pi_S \pi_Q \pi_E \ \frac{\text{failures}}{10^6 \ \text{hours}} \ .$$

Base Failure Rate - λ_b

Type	λ_b
NPN or PNP	0.00074

Temperature Factor - π_T

$$\pi_T = \exp\left(-2114\left(\frac{1}{T_J + 273} - \frac{1}{298} \right) \right)$$

Application Factor - π_A

Application	π_A
Linear Amplification	1.50
Switching	0.70

Power Rating Factor - π_R

$\pi_R = 0.43$, if $P_R \le 0.1W$

$\pi_R = P_R^{0.37}$, if $P_R > 0.1W$

where $P_R = $ Rated Power

Voltage Stress Factor - π_S

$\pi_S = 0.045 \exp(3.1V_s), \quad 0 \le V_s \le 1$

$V_s = \dfrac{\text{applied } V_{CE}}{\text{rated } V_{CEO}}$

$V_{CE} = $ voltage, collector to emitter

$V_{CEO} = $ voltage, collector to emitter, base open

Part Quality Factor - π_Q

Quality	π_Q
JANTXV	0.7
JANTX	1.0
JAN	2.4
Lower	5.5
Plastic	8.0

Environmental Factor - π_E

Environment	π_E
G$_B$	1
G$_F$	6
G$_M$	9
N$_S$	9
N$_U$	19
A$_{IC}$	13
A$_{IF}$	29
A$_{UC}$	20
A$_{UF}$	43
A$_{RW}$	24
S$_F$	0.50
M$_F$	14
M$_L$	32
C$_L$	320

B.3.4 Transistors: Bipolar Junction, High Frequency (> 200MHz), Low Noise (Power ≤ 1W)

The failure rate is given by the following relationship

$$\lambda = \lambda_b \pi_T \pi_R \pi_S \pi_Q \pi_E \ \frac{\text{failures}}{10^6 \text{ hours}} \ .$$

Base Failure Rate - λ_b

Type	λ_b
All types	0.18

Power Rating Factor - π_R

$\pi_R = 0.43$, if $P_R \le 0.1\text{W}$

$\pi_R = P_R^{0.37}$, if $P_R > 0.1\text{W}$

where P_R = Rated Power

Part Quality Factor - π_Q

Quality	π_Q
JANTXV	0.5
JANTX	1.0
JAN	2.0
Lower	5.0

Temperature Factor - π_T

$$\pi_T = \exp\left(-2114\left(\frac{1}{T_J + 273} - \frac{1}{298}\right)\right)$$

Voltage Stress Factor - π_S

$\pi_s = 0.045 \exp(3.1 V_s), \quad 0 \le V_s \le 1$

$V_s = \dfrac{\text{applied } V_{CE}}{\text{rated } V_{CEO}}$

V_{CE} = voltage, collector to emitter

V_{CEO} = voltage, collector to emitter, base open

Environmental Factor - π_E

Environment	π_E
G$_B$	1
G$_F$	2
G$_M$	5
N$_S$	4
N$_U$	11
A$_{IC}$	4
A$_{IF}$	5
A$_{UC}$	7
A$_{UF}$	12
A$_{RW}$	16
S$_F$	0.50
M$_F$	9
M$_L$	24
C$_L$	250

B.3.5 Transistors: Field Effect, Low Frequency (≤ 400MHz)

The failure rate is given by the following relationship

$$\lambda = \lambda_b \pi_T \pi_A \pi_Q \pi_E \frac{\text{failures}}{10^6 \text{ hours}}.$$

Base Failure Rate - λ_b

Type	λ_b
MOSFET	0.012
JFET	0.0045

Temperature Factor - π_T

$$\pi_T = \exp\left(-1925\left(\frac{1}{T_J + 273} - \frac{1}{298}\right)\right)$$

Application Factor - π_A

Application (P_R = rated output power)	π_A
Linear Amplification ($P_R < 2W$)	1.50
Small Signal Switching	0.70
Power FETs (Non-linear, $P_R \geq 2W$)	
$2 \leq P_R < 5W$	2.0
$5 \leq P_R < 50W$	4.0
$50 \leq P_R < 250W$	8.0
$P_R \geq 250W$	10.0

Part Quality Factor - π_Q

Quality	π_Q
JANTXV	0.7
JANTX	1.0
JAN	2.4
Lower	5.5
Plastic	8.0

Environmental Factor - π_E

Environment	π_E
G$_B$	1
G$_F$	6
G$_M$	9
N$_S$	9
N$_U$	19
A$_{IC}$	13
A$_{IF}$	29
A$_{UC}$	20
A$_{UF}$	43
A$_{RW}$	24
S$_F$	0.50
M$_F$	14
M$_L$	32
C$_L$	320

B.3.6 Transistors: Field Effect, High Frequency (> 400MHz), Low Power (≤ 300mW)

The failure rate is given by the following relationship

$$\lambda = \lambda_b \pi_T \pi_Q \pi_E \ \frac{\text{failures}}{10^6 \ \text{hours}} \ .$$

Base Failure Rate - λ_b

Type	λ_b
MOSFET	0.060
JFET	0.023

Part Quality Factor - π_Q

Quality	π_Q
JANTXV	0.5
JANTX	1.0
JAN	2.0
Lower	5.0

Temperature Factor - π_T

$$\pi_T = \exp\left(-1925\left(\frac{1}{T_J + 273} - \frac{1}{298} \right) \right)$$

Environmental Factor - π_E

Environment	π_E
G_B	1
G_F	2
G_M	5
N_S	4
N_U	11
A_{IC}	4
A_{IF}	5
A_{UC}	7
A_{UF}	12
A_{RW}	16
S_F	0.50
M_F	9
M_L	24
C_L	250

B.3.7 Microcircuits: Gate/Logic Arrays and Microprocessors

Includes the following devices:

1. Bipolar devices, Digital and Linear Gate/Logic Arrays

2. MOS Devices, Digital and Linear Gate/Logic Arrays

3. Field Programmable Logic Array (PLA) and Programmable Array Logic (PAL)

4. Microprocessors

The failure rate is given by the following relationship

$$\lambda = \left(C_1\pi_T + C_2\pi_E\right)\pi_Q\pi_L \frac{\text{failures}}{10^6 \text{ hours}} \ .$$

C_1 = Complexity Failure Rate for Bipolar Devices (Digital and Linear Gate/Logic)

Digital		Linear		PLA/PAL	
No. Gates	C_1	No. Transistors	C_1	No. Gates	C_1
1 to 100	0.0025	1 to 100	0.010	Up to 200	0.010
101 to 1,000	0.0050	101 to 300	0.020	201 to 1,000	0.021
1,001 to 3,000	0.010	301 to 1,000	0.040	1,001 to 5,000	0.042
3,001 to 10,000	0.020	1,001 to 10,000	0.060		
10,001 to 30,000	0.040				
30,001 to 60,000	0.080				

C_1 = Complexity Failure Rate for MOS Devices (Digital and Linear Gate/Logic)

Digital		Linear		MOS	
No. Gates	C_1	No. Transistors	C_1	No. Gates	C_1
1 to 100	0.010	1 to 100	0.010	Up to 500	0.00085
101 to 1,000	0.020	101 to 300	0.020	501 to 1,000	0.0017
1,001 to 3,000	0.040	301 to 1,000	0.040	1,001 to 5,000	0.0034
3,001 to 10,000	0.080	1,001 to 10,000	0.060	5,001 to 20,000	0.0068
10,001 to 30,000	0.16				
30,001 to 60,000	0.29				

$$C_1 = \textbf{Complexity Failure Rate for Microprocessors}$$

Number of Bits	C_1 - Bipolar	C_1 - MOS
Up to 8	0.060	0.14
Up to 16	0.12	0.28
Up to 32	0.24	0.56

C_2 = **Package Failure Rate for all microcircuits.** N_p = **number of pins on package.**

$C_2 = 2.8 \times 10^{-4} (N_p)^{1.08}$, Hermetic : DIPs w/ solder or weld seal, SMT

$C_2 = 9.0 \times 10^{-5} (N_p)^{1.51}$, DIPs with glass seal

$C_2 = 3.0 \times 10^{-5} (N_p)^{1.82}$, Flatpacks with axial leads on 50 mil centers

$C_2 = 2.8 \times 10^{-4} (N_p)^{2.01}$, Cans

$C_2 = 3.0 \times 10^{-5} (N_p)^{1.08}$, Nonhermetic : DIPs, PGA, SMT (leaded and nonleaded)

Temperature Factor - π_T

$$\pi_T = 0.1 \times \exp\left(-\frac{E_a}{8.617 \times 10^{-5}} \left(\frac{1}{T_J + 273} - \frac{1}{296} \right) \right)$$

The activation energy (E_a) for different technologies are given in the table below.

Technology	TTL, ASTLL, CML, HTTL, FTLL, DTL, ECL, ALSTTL	F, LTTL, STTL	BiCMOS LSTTL	Digital MOS, VHSIC CMOS	Linear (Bipolar and MOS)	Memories (Bipolar and MOS), NMOS
E_a	0.4	0.45	0.5	0.35	0.65	0.6

Learning Factor - π_L

Years in Production	π_L
$\le .1$	2.0
0.5	1.8
1.0	1.5
1.5	1.2
≥ 2.0	1.0

Part Quality Factor - π_Q

Quality	π_Q
S	0.25
B	1.0
B-1	2.0

Environmental Factor - π_E

Environment	π_E
G_B	0.50
G_F	2
G_M	4
N_S	4
N_U	6
A_{IC}	4
A_{IF}	5
A_{UC}	5
A_{UF}	8
A_{RW}	8
S_F	0.50
M_F	5
M_L	12
C_L	220

Appendix C Manufacturer Datasheets

This appendix contains manufacturer datasheets for selected electronic components in support of Chapter 8. The following datasheets are included:

- 1N4001 Rectifier Diode.
- 2N3904 General Purpose Amplifier NPN Transistor.
- CD4001BM/BC Quad 2-Input NOR Gate.
- LM741 General Purpose Operational Amplifier.

1N4001 - 1N4007

Features

* Low forward voltage drop.

* High surge current capability.

DO-41
COLOR BAND DENOTES CATHODE

General Purpose Rectifiers (Glass Passivated)

Absolute Maximum Ratings* $T_A = 25°C$ unless otherwise noted

Symbol	Parameter	Value							Units
		4001	4002	4003	4004	4005	4006	4007	
V_{RRM}	Peak Repetitive Reverse Voltage	50	100	200	400	600	800	1000	V
$I_{F(AV)}$	Average Rectified Forward Current, .375 " lead length @ $T_A = 75°C$	1.0							A
I_{FSM}	Non-repetitive Peak Forward Surge Current 8.3 ms Single Half-Sine-Wave	30							A
T_{stg}	Storage Temperature Range	-55 to +175							°C
T_J	Operating Junction Temperature	-55 to +175							°C

*These ratings are limiting values above which the serviceability of any semiconductor device may be impaired.

Thermal Characteristics

Symbol	Parameter	Value	Units
P_D	Power Dissipation	3.0	W
$R_{θJA}$	Thermal Resistance, Junction to Ambient	50	°C/W

Electrical Characteristics $T_A = 25°C$ unless otherwise noted

Symbol	Parameter	Device							Units
		4001	4002	4003	4004	4005	4006	4007	
V_F	Forward Voltage @ 1.0 A	1.1							V
I_{rr}	Maximum Full Load Reverse Current, Full Cycle $T_A = 75°C$	30							μA
I_R	Reverse Current @ rated V_R $T_A = 25°C$ $T_A = 100°C$	5.0 500							μA μA
C_T	Total Capacitance $V_R = 4.0$ V, f = 1.0 MHz	15							pF

2N3904 / MMBT3904 / MMPQ3904 / PZT3904

Discrete Power & Signal
Technologies

2N3904

TO-92

MMBT3904

SOT-23
Mark: 1A

MMPQ3904

SOIC-16

PZT3904

SOT-223

NPN General Purpose Amplifier

This device is designed as a general purpose amplifier and switch.
The useful dynamic range extends to 100 mA as a switch and to
100 MHz as an amplifier. Sourced from Process 23.

Absolute Maximum Ratings* TA = 25°C unless otherwise noted

Symbol	Parameter	Value	Units
V_{CEO}	Collector-Emitter Voltage	40	V
V_{CBO}	Collector-Base Voltage	60	V
V_{EBO}	Emitter-Base Voltage	6.0	V
I_C	Collector Current - Continuous	200	mA
T_J, T_{stg}	Operating and Storage Junction Temperature Range	-55 to +150	°C

*These ratings are limiting values above which the serviceability of any semiconductor device may be impaired.

NOTES:
1) These ratings are based on a maximum junction temperature of 150 degrees C.
2) These are steady state limits. The factory should be consulted on applications involving pulsed or low duty cycle operations.

NPN General Purpose Amplifier
(continued)

Thermal Characteristics   TA = 25°C unless otherwise noted

Symbol	Characteristic	Max		Units
		2N3904	*PZT3904	
P_D	Total Device Dissipation	625	1,000	mW
	Derate above 25°C	5.0	8.0	mW/°C
$R_{\theta JC}$	Thermal Resistance, Junction to Case	83.3		°C/W
$R_{\theta JA}$	Thermal Resistance, Junction to Ambient	200	125	°C/W

Symbol	Characteristic	Max		Units
		**MMBT3904	MMPQ3904	
P_D	Total Device Dissipation	350	1,000	mW
	Derate above 25°C	2.8	8.0	mW/°C
$R_{\theta JA}$	Thermal Resistance, Junction to Ambient	357		°C/W
	Effective 4 Die		125	°C/W
	Each Die		240	°C/W

*Device mounted on FR-4 PCB 36 mm X 18 mm X 1.5 mm; mounting pad for the collector lead min. 6 cm².

**Device mounted on FR-4 PCB 1.6" X 1.6" X 0.06."

Typical Characteristics

Typical Pulsed Current Gain vs Collector Current

Collector-Emitter Saturation Voltage vs Collector Current

Base-Emitter Saturation Voltage vs Collector Current

Base-Emitter ON Voltage vs Collector Current

National Semiconductor

March 1988

CD4001BM/CD4001BC Quad 2-Input NOR Buffered B Series Gate
CD4011BM/CD4011BC Quad 2-Input NAND Buffered B Series Gate

General Description

These quad gates are monolithic complementary MOS (CMOS) integrated circuits constructed with N- and P-channel enhancement mode transistors. They have equal source and sink current capabilities and conform to standard B series output drive. The devices also have buffered outputs which improve transfer characteristics by providing very high gain.

All inputs are protected against static discharge with diodes to V_{DD} and V_{SS}.

Features

- Low power TTL compatibility — Fan out of 2 driving 74L or 1 driving 74LS
- 5V–10V–15V parametric ratings
- Symmetrical output characteristics
- Maximum input leakage 1 μA at 15V over full temperature range

Schematic Diagrams

CD4001BC/BM

1/4 of device shown
$J = \overline{A + B}$
Logical "1" = High
Logical "0" = Low

TL/F/5939–2

*All inputs protected by standard CMOS protection circuit.

TL/F/5939–1

CD4011BC/BM

1/4 of device shown
$J = \overline{A \cdot B}$
Logical "1" = High
Logical "0" = Low

TL/F/5939–6

*All inputs protected by standard CMOS protection circuit.

TL/F/5939–5

RRD-B30M105/Printed in U. S. A.

Absolute Maximum Ratings (Notes 1 and 2)

If Military/Aerospace specified devices are required, please contact the National Semiconductor Sales Office/Distributors for availability and specifications.

Voltage at any Pin	$-0.5V$ to $V_{DD} + 0.5V$
Power Dissipation (P_D)	
Dual-In-Line	700 mW
Small Outline	500 mW
V_{DD} Range	$-0.5 V_{DC}$ to $+18 V_{DC}$
Storage Temperature (T_S)	$-65°C$ to $+150°C$
Lead Temperature (T_L)	
(Soldering, 10 seconds)	260°C

Operating Conditions

Operating Range (V_{DD})	$3 V_{DC}$ to $15 V_{DC}$
Operating Temperature Range	
CD4001BM, CD4011BM	$-55°C$ to $+125°C$
CD4001BC, CD4011BC	$-40°C$ to $+85°C$

DC Electrical Characteristics CD4001BM, CD4011BM (Note 2)

Symbol	Parameter	Conditions	$-55°C$ Min	$-55°C$ Max	$+25°C$ Min	$+25°C$ Typ	$+25°C$ Max	$+125°C$ Min	$+125°C$ Max	Units
I_{DD}	Quiescent Device Current	$V_{DD} = 5V$, $V_{IN} = V_{DD}$ or V_{SS}		0.25		0.004	0.25		7.5	μA
		$V_{DD} = 10V$, $V_{IN} = V_{DD}$ or V_{SS}		0.50		0.005	0.50		15	μA
		$V_{DD} = 15V$, $V_{IN} = V_{DD}$ or V_{SS}		1.0		0.006	1.0		30	μA
V_{OL}	Low Level Output Voltage	$V_{DD} = 5V$		0.05		0	0.05		0.05	V
		$V_{DD} = 10V$ $\}$ $\|I_O\| < 1$ μA		0.05		0	0.05		0.05	V
		$V_{DD} = 15V$		0.05		0	0.05		0.05	V
V_{OH}	High Level Output Voltage	$V_{DD} = 5V$	4.95		4.95	5		4.95		V
		$V_{DD} = 10V$ $\}$ $\|I_O\| < 1$ μA	9.95		9.95	10		9.95		V
		$V_{DD} = 15V$	14.95		14.95	15		14.95		V
V_{IL}	Low Level Input Voltage	$V_{DD} = 5V$, $V_O = 4.5V$		1.5		2	1.5		1.5	V
		$V_{DD} = 10V$, $V_O = 9.0V$		3.0		4	3.0		3.0	V
		$V_{DD} = 15V$, $V_O = 13.5V$		4.0		6	4.0		4.0	V
V_{IH}	High Level Input Voltage	$V_{DD} = 5V$, $V_O = 0.5V$	3.5		3.5	3		3.5		V
		$V_{DD} = 10V$, $V_O = 1.0V$	7.0		7.0	6		7.0		V
		$V_{DD} = 15V$, $V_O = 1.5V$	11.0		11.0	9		11.0		V
I_{OL}	Low Level Output Current (Note 3)	$V_{DD} = 5V$, $V_O = 0.4V$	0.64		0.51	0.88		0.36		mA
		$V_{DD} = 10V$, $V_O = 0.5V$	1.6		1.3	2.25		0.9		mA
		$V_{DD} = 15V$, $V_O = 1.5V$	4.2		3.4	8.8		2.4		mA
I_{OH}	High Level Output Current (Note 3)	$V_{DD} = 5V$, $V_O = 4.6V$	-0.64		-0.51	-0.88		-0.36		mA
		$V_{DD} = 10V$, $V_O = 9.5V$	-1.6		-1.3	-2.25		-0.9		mA
		$V_{DD} = 15V$, $V_O = 13.5V$	-4.2		-3.4	-8.8		-2.4		mA
I_{IN}	Input Current	$V_{DD} = 15V$, $V_{IN} = 0V$		-0.10		-10^{-5}	-0.10		-1.0	μA
		$V_{DD} = 15V$, $V_{IN} = 15V$		0.10		10^{-5}	0.10		1.0	μA

Connection Diagrams

CD4001BC/CD4001BM
Dual-In-Line Package

TL/F/5939-3

Top View

CD4011BC/CD4011BM
Dual-In-Line Package

TL/F/5939-4

Top View
Order Number CD4001B or CD4011B

2

DC Electrical Characteristics CD4001BC, CD4011BC (Note 2)

Symbol	Parameter	Conditions	−40°C Min	−40°C Max	+25°C Min	+25°C Typ	+25°C Max	+85°C Min	+85°C Max	Units
I_{DD}	Quiescent Device Current	$V_{DD}=5V$, $V_{IN}=V_{DD}$ or V_{SS}		1		0.004	1		7.5	µA
		$V_{DD}=10V$, $V_{IN}=V_{DD}$ or V_{SS}		2		0.005	2		15	µA
		$V_{DD}=15V$, $V_{IN}=V_{DD}$ or V_{SS}		4		0.006	4		30	µA
V_{OL}	Low Level Output Voltage	$V_{DD}=5V$		0.05		0	0.05		0.05	V
		$V_{DD}=10V$ $\}$ $|I_O|<1$ µA		0.05		0	0.05		0.05	V
		$V_{DD}=15V$		0.05		0	0.05		0.05	V
V_{OH}	High Level Output Voltage	$V_{DD}=5V$	4.95		4.95	5		4.95		V
		$V_{DD}=10V$ $\}$ $|I_O|<1$ µA	9.95		9.95	10		9.95		V
		$V_{DD}=15V$	14.95		14.95	15		14.95		V
V_{IL}	Low Level Input Voltage	$V_{DD}=5V$, $V_O=4.5V$		1.5		2	1.5		1.5	V
		$V_{DD}=10V$, $V_O=9.0V$		3.0		4	3.0		3.0	V
		$V_{DD}=15V$, $V_O=13.5V$		4.0		6	4.0		4.0	V
V_{IH}	High Level Input Voltage	$V_{DD}=5V$, $V_O=0.5V$	3.5		3.5	3		3.5		V
		$V_{DD}=10V$, $V_O=1.0V$	7.0		7.0	6		7.0		V
		$V_{DD}=15V$, $V_O=1.5V$	11.0		11.0	9		11.0		V
I_{OL}	Low Level Output Current (Note 3)	$V_{DD}=5V$, $V_O=0.4V$	0.52		0.44	0.88		0.36		mA
		$V_{DD}=10V$, $V_O=0.5V$	1.3		1.1	2.25		0.9		mA
		$V_{DD}=15V$, $V_O=1.5V$	3.6		3.0	8.8		2.4		mA
I_{OH}	High Level Output Current (Note 3)	$V_{DD}=5V$, $V_O=4.6V$	−0.52		−0.44	−0.88		−0.36		mA
		$V_{DD}=10V$, $V_O=9.5V$	−1.3		−1.1	−2.25		−0.9		mA
		$V_{DD}=15V$, $V_O=13.5V$	−3.6		−3.0	−8.8		−2.4		mA
I_{IN}	Input Current	$V_{DD}=15V$, $V_{IN}=0V$		−0.30		-10^{-5}	−0.30		−1.0	µA
		$V_{DD}=15V$, $V_{IN}=15V$		0.30		10^{-5}	0.30		1.0	µA

AC Electrical Characteristics* CD4001BC, CD4001BM

$T_A=25°C$, Input t_r, $t_f=20$ ns, $C_L=50$ pF, $R_L=200k$. Typical temperature coefficient is 0.3%/°C.

Symbol	Parameter	Conditions	Typ	Max	Units
t_{PHL}	Propagation Delay Time, High-to-Low Level	$V_{DD}=5V$	120	250	ns
		$V_{DD}=10V$	50	100	ns
		$V_{DD}=15V$	35	70	ns
t_{PLH}	Propagation Delay Time, Low-to-High Level	$V_{DD}=5V$	110	250	ns
		$V_{DD}=10V$	50	100	ns
		$V_{DD}=15V$	35	70	ns
t_{THL}, t_{TLH}	Transition Time	$V_{DD}=5V$	90	200	ns
		$V_{DD}=10V$	50	100	ns
		$V_{DD}=15V$	40	80	ns
C_{IN}	Average Input Capacitance	Any Input	5	7.5	pF
C_{PD}	Power Dissipation Capacity	Any Gate	14		pF

*AC Parameters are guaranteed by DC correlated testing.

Note 1: "Absolute Maximum Ratings" are those values beyond which the safety of the device cannot be guaranteed. Except for "Operating Temperature Range" they are not meant to imply that the devices should be operated at these limits. The table of "Electrical Characteristics" provides conditions for actual device operation.

Note 2: All voltages measured with respect to V_{SS} unless otherwise specified.

Note 3: I_{OL} and I_{OH} are tested one output at a time.

3

National Semiconductor

November 1994

LM741 Operational Amplifier

General Description

The LM741 series are general purpose operational amplifiers which feature improved performance over industry standards like the LM709. They are direct, plug-in replacements for the 709C, LM201, MC1439 and 748 in most applications.

The amplifiers offer many features which make their application nearly foolproof: overload protection on the input and output, no latch-up when the common mode range is exceeded, as well as freedom from oscillations.

The LM741C/LM741E are identical to the LM741/LM741A except that the LM741C/LM741E have their performance guaranteed over a 0°C to +70°C temperature range, instead of −55°C to +125°C.

Schematic Diagram

TL/H/9341–1

Offset Nulling Circuit

TL/H/9341–7

Absolute Maximum Ratings

If Military/Aerospace specified devices are required, please contact the National Semiconductor Sales Office/ Distributors for availability and specifications. (Note 5)

	LM741A	LM741E	LM741	LM741C
Supply Voltage	±22V	±22V	±22V	±18V
Power Dissipation (Note 1)	500 mW	500 mW	500 mW	500 mW
Differential Input Voltage	±30V	±30V	±30V	±30V
Input Voltage (Note 2)	±15V	±15V	±15V	±15V
Output Short Circuit Duration	Continuous	Continuous	Continuous	Continuous
Operating Temperature Range	−55°C to +125°C	0°C to +70°C	−55°C to +125°C	0°C to +70°C
Storage Temperature Range	−65°C to +150°C	−65°C to +150°C	−65°C to +150°C	−65°C to +150°C
Junction Temperature	150°C	100°C	150°C	100°C
Soldering Information				
N-Package (10 seconds)	260°C	260°C	260°C	260°C
J- or H-Package (10 seconds)	300°C	300°C	300°C	300°C
M-Package				
Vapor Phase (60 seconds)	215°C	215°C	215°C	215°C
Infrared (15 seconds)	215°C	215°C	215°C	215°C

See AN-450 "Surface Mounting Methods and Their Effect on Product Reliability" for other methods of soldering surface mount devices.

ESD Tolerance (Note 6)	400V	400V	400V	400V

Electrical Characteristics (Note 3)

Parameter	Conditions	LM741A/LM741E			LM741			LM741C			Units
		Min	Typ	Max	Min	Typ	Max	Min	Typ	Max	
Input Offset Voltage	$T_A = 25°C$ $R_S \leq 10\ k\Omega$ $R_S \leq 50\Omega$		0.8	3.0		1.0	5.0		2.0	6.0	mV mV
	$T_{AMIN} \leq T_A \leq T_{AMAX}$ $R_S \leq 50\Omega$ $R_S \leq 10\ k\Omega$			4.0			6.0			7.5	mV mV
Average Input Offset Voltage Drift				15							μV/°C
Input Offset Voltage Adjustment Range	$T_A = 25°C,\ V_S = \pm20V$	±10				±15			±15		mV
Input Offset Current	$T_A = 25°C$		3.0	30		20	200		20	200	nA
	$T_{AMIN} \leq T_A \leq T_{AMAX}$			70		85	500			300	nA
Average Input Offset Current Drift				0.5							nA/°C
Input Bias Current	$T_A = 25°C$		30	80		80	500		80	500	nA
	$T_{AMIN} \leq T_A \leq T_{AMAX}$			0.210			1.5			0.8	μA
Input Resistance	$T_A = 25°C,\ V_S = \pm20V$	1.0	6.0		0.3	2.0		0.3	2.0		MΩ
	$T_{AMIN} \leq T_A \leq T_{AMAX}$, $V_S = \pm20V$	0.5									MΩ
Input Voltage Range	$T_A = 25°C$							±12	±13		V
	$T_{AMIN} \leq T_A \leq T_{AMAX}$				±12	±13					V
Large Signal Voltage Gain	$T_A = 25°C,\ R_L \geq 2\ k\Omega$ $V_S = \pm20V,\ V_O = \pm15V$ $V_S = \pm15V,\ V_O = \pm10V$	50			50	200		20	200		V/mV V/mV
	$T_{AMIN} \leq T_A \leq T_{AMAX}$, $R_L \geq 2\ k\Omega$, $V_S = \pm20V,\ V_O = \pm15V$ $V_S = \pm15V,\ V_O = \pm10V$ $V_S = \pm5V,\ V_O = \pm2V$	32 10			25			15			V/mV V/mV V/mV

2

Electrical Characteristics (Note 3) (Continued)

Parameter	Conditions	LM741A/LM741E			LM741			LM741C			Units
		Min	Typ	Max	Min	Typ	Max	Min	Typ	Max	
Output Voltage Swing	$V_S = \pm 20V$										
	$R_L \geq 10\,k\Omega$	±16									V
	$R_L \geq 2\,k\Omega$	±15									V
	$V_S = \pm 15V$										
	$R_L \geq 10\,k\Omega$				±12	±14		±12	±14		V
	$R_L \geq 2\,k\Omega$				±10	±13		±10	±13		V
Output Short Circuit Current	$T_A = 25°C$	10	25	35		25			25		mA
	$T_{AMIN} \leq T_A \leq T_{AMAX}$	10		40							mA
Common-Mode Rejection Ratio	$T_{AMIN} \leq T_A \leq T_{AMAX}$										
	$R_S \leq 10\,k\Omega$, $V_{CM} = \pm 12V$				70	90		70	90		dB
	$R_S \leq 50\Omega$, $V_{CM} = \pm 12V$	80	95								dB
Supply Voltage Rejection Ratio	$T_{AMIN} \leq T_A \leq T_{AMAX}$,										
	$V_S = \pm 20V$ to $V_S = \pm 5V$										
	$R_S \leq 50\Omega$	86	96								dB
	$R_S \leq 10\,k\Omega$				77	96		77	96		dB
Transient Response	$T_A = 25°C$, Unity Gain										
Rise Time			0.25	0.8		0.3			0.3		µs
Overshoot			6.0	20		5			5		%
Bandwidth (Note 4)	$T_A = 25°C$	0.437	1.5								MHz
Slew Rate	$T_A = 25°C$, Unity Gain	0.3	0.7			0.5			0.5		V/µs
Supply Current	$T_A = 25°C$					1.7	2.8		1.7	2.8	mA
Power Consumption	$T_A = 25°C$										
	$V_S = \pm 20V$		80	150							mW
	$V_S = \pm 15V$					50	85		50	85	mW
LM741A	$V_S = \pm 20V$										
	$T_A = T_{AMIN}$			165							mW
	$T_A = T_{AMAX}$			135							mW
LM741E	$V_S = \pm 20V$										
	$T_A = T_{AMIN}$			150							mW
	$T_A = T_{AMAX}$			150							mW
LM741	$V_S = \pm 15V$										
	$T_A = T_{AMIN}$					60	100				mW
	$T_A = T_{AMAX}$					45	75				mW

Note 1: For operation at elevated temperatures, these devices must be derated based on thermal resistance, and T_j max. (listed under "Absolute Maximum Ratings"). $T_j = T_A + (\theta_{jA} P_D)$.

Thermal Resistance	Cerdip (J)	DIP (N)	HO8 (H)	SO-8 (M)
θ_{jA} (Junction to Ambient)	100°C/W	100°C/W	170°C/W	195°C/W
θ_{jC} (Junction to Case)	N/A	N/A	25°C/W	N/A

Note 2: For supply voltages less than ±15V, the absolute maximum input voltage is equal to the supply voltage.

Note 3: Unless otherwise specified, these specifications apply for $V_S = \pm 15V$, $-55°C \leq T_A \leq +125°C$ (LM741/LM741A). For the LM741C/LM741E, these specifications are limited to $0°C \leq T_A \leq +70°C$.

Note 4: Calculated value from: BW (MHz) = 0.35/Rise Time(µs).

Note 5: For military specifications see RETS741X for LM741 and RETS741AX for LM741A.

Note 6: Human body model, 1.5 kΩ in series with 100 pF.

3

Index

About the Authors

Ralph Ford obtained his Ph.D. and M.S. degrees in Electrical Engineering from the University of Arizona in 1994 and 1989 respectively. He obtained his B.S. in Electrical Engineering from Clarkson University in 1987. He worked for the IBM Microelectronics Division in East Fishkill, NY from 1989-1991, where he developed machine vision systems to inspect electronic packaging modules for mainframe computers. Ralph also has experience working for IBM Data Systems and the Brookhaven National Laboratory. He joined the faculty at Penn State Erie, The Behrend College in 1994. Ralph has experience teaching electronics and software design, as well as teaching the capstone design course sequence in the electrical, computer, and software engineering programs. His research interests are in engineering design, image processing, machine vision, and signal processing. Ralph is Program Chair of Electrical, Computer, and Software Engineering at Penn State Behrend, and also serves as a program evaluator for ABET. He was awarded a Fulbright Scholarship to study at the Brno University of Technology in the Czech Republic in 2005.

Chris Coulston obtained his Ph.D. in Computer Science and Engineering from the Pennsylvania State University in 1999. He obtained his M.S. and B.S in Computer Engineering from the Pennsylvania State University in 1994 and 1992 respectively. Chris has industry experience working for IBM in Manassas, VA and Accu-Weather in State College, PA. He joined the faculty at Penn State Erie, The Behrend College in 1999. He has experience teaching design-oriented courses in digital systems, embedded systems, computer architecture, and database management systems. Chris' research interests are in Steiner tree routing algorithms and artificial life. He is currently an assistant professor in Electrical, Computer and Software Engineering at Penn State Behrend.